Student Edition

Eureka Math
Algebra II
Modules 1 & 2

Special thanks go to the Gordan A. Cain Center and to the Department of Mathematics at Louisiana State University for their support in the development of *Eureka Math*.

Published by Great Minds

Printed in the U.S.A.
This book may be purchased from the publisher at eureka-math.org
10 9 8 7 6 5 4 3

ISBN 978-1-63255-330-0

Lesson 1: Successive Differences in Polynomials

Classwork

Opening Exercise

John noticed patterns in the arrangement of numbers in the table below.

________	2.4		3.4		4.4		5.4		6.4
________	5.76		11.56		19.36		29.16		40.96
________		5.8		7.8		9.8		11.8	
________			2		2		2		

Assuming that the pattern would continue, he used it to find the value of 7.4^2. Explain how he used the pattern to find 7.4^2, and then use the pattern to find 8.4^2.

How would you label each row of numbers in the table?

Discussion

Let the sequence $\{a_0, a_1, a_2, a_3, \ldots\}$ be generated by evaluating a polynomial expression at the values $0, 1, 2, 3, \ldots$ The numbers found by evaluating $a_1 - a_0$, $a_2 - a_1$, $a_3 - a_2$, ... form a new sequence, which we will call the *first differences* of the polynomial. The differences between successive terms of the first differences sequence are called the *second differences*, and so on.

Example 1

What is the sequence of first differences for the linear polynomial given by $ax + b$, where a and b are constant coefficients?

What is the sequence of second differences for $ax + b$?

Example 2

Find the first, second, and third differences of the polynomial $ax^2 + bx + c$ by filling in the blanks in the following table.

x	$ax^2 + bx + c$	First Differences	Second Differences	Third Differences
0	c			
1	$a + b + c$			
2	$4a + 2b + c$			
3	$9a + 3b + c$			
4	$16a + 4b + c$			
5	$25a + 5b + c$			

Example 3

Find the second, third, and fourth differences of the polynomial $ax^3 + bx^2 + cx + d$ by filling in the blanks in the following table.

x	$ax^3 + bx^2 + cx + d$	First Differences	Second Differences	Third Differences	Fourth Differences
0	d				
		$a + b + c$			
1	$a + b + c + d$				
		$7a + 3b + c$			
2	$8a + 4b + 2c + d$				
		$19a + 5b + c$			
3	$27a + 9b + 3c + d$				
		$37a + 7b + c$			
4	$64a + 16b + 4c + d$				
		$61a + 9b + c$			
5	$125a + 25b + 5c + d$				

Example 4

What type of relationship does the set of ordered pairs (x, y) satisfy? How do you know? Fill in the blanks in the table below to help you decide. (The first differences have already been computed for you.)

x	y	First Differences	Second Differences	Third Differences
0	2			
		-1		
1	1			
		5		
2	6			
		17		
3	23			
		35		
4	58			
		59		
5	117			

Find the equation of the form $y = ax^3 + bx^2 + cx + d$ that all ordered pairs (x, y) above satisfy. Give evidence that your equation is correct.

Relevant Vocabulary

NUMERICAL SYMBOL: A *numerical symbol* is a symbol that represents a specific number. Examples: $1, 2, 3, 4, \pi, -3.2$.

VARIABLE SYMBOL: A *variable symbol* is a symbol that is a placeholder for a number from a specified set of numbers. The set of numbers is called the *domain of the variable*. Examples: x, y, z.

ALGEBRAIC EXPRESSION: An *algebraic expression* is either

1. a numerical symbol or a variable symbol or
2. the result of placing previously generated algebraic expressions into the two blanks of one of the four operators ((__)+(__), (__)−(__), (__)×(__), (__)÷(__)) or into the base blank of an exponentiation with an exponent that is a rational number.

Following the definition above, $\big(((x) \times (x)) \times (x)\big) + ((3) \times (x))$ is an algebraic expression, but it is generally written more simply as $x^3 + 3x$.

NUMERICAL EXPRESSION: A *numerical expression* is an algebraic expression that contains only numerical symbols (no variable symbols) that evaluates to a single number. Example: The numerical expression $\frac{(3 \cdot 2)^2}{12}$ evaluates to 3.

MONOMIAL: A *monomial* is an algebraic expression generated using only the multiplication operator (__×__). The expressions x^3 and $3x$ are both monomials.

BINOMIAL: A *binomial* is the sum of two monomials. The expression $x^3 + 3x$ is a binomial.

POLYNOMIAL EXPRESSION: A *polynomial expression* is a monomial or sum of two or more monomials.

SEQUENCE: A *sequence* can be thought of as an ordered list of elements. The elements of the list are called the *terms of the sequence.*

ARITHMETIC SEQUENCE: A sequence is called *arithmetic* if there is a real number d such that each term in the sequence is the sum of the previous term and d.

Problem Set

1. Create a table to find the second differences for the polynomial $36 - 16t^2$ for integer values of t from 0 to 5.

2. Create a table to find the third differences for the polynomial $s^3 - s^2 + s$ for integer values of s from -3 to 3.

3. Create a table of values for the polynomial x^2, using $n, n+1, n+2, n+3, n+4$ as values of x. Show that the second differences are all equal to 2.

4. Show that the set of ordered pairs (x, y) in the table below satisfies a quadratic relationship. (Hint: Find second differences.) Find the equation of the form $y = ax^2 + bx + c$ that all of the ordered pairs satisfy.

x	0	1	2	3	4	5
y	5	4	-1	-10	-23	-40

5. Show that the set of ordered pairs (x, y) in the table below satisfies a cubic relationship. (Hint: Find third differences.) Find the equation of the form $y = ax^3 + bx^2 + cx + d$ that all of the ordered pairs satisfy.

x	0	1	2	3	4	5
y	20	4	0	20	76	180

6. The distance d ft. required to stop a car traveling at $10v$ mph under dry asphalt conditions is given by the following table.

v	0	1	2	3	4	5
d	0	5	19.5	43.5	77	120

   a. What type of relationship is indicated by the set of ordered pairs?

   b. Assuming that the relationship continues to hold, find the distance required to stop the car when the speed reaches 60 mph, when $v = 6$.

   c. Extension: Find an equation that describes the relationship between the speed of the car v and its stopping distance d.

7. Use the polynomial expressions $5x^2 + x + 1$ and $2x + 3$ to answer the questions below.

   a. Create a table of second differences for the polynomial $5x^2 + x + 1$ for the integer values of x from 0 to 5.

   b. Justin claims that for $n \geq 2$, the nth differences of the sum of a degree n polynomial and a linear polynomial are the same as the nth differences of just the degree n polynomial. Find the second differences for the sum $(5x^2 + x + 1) + (2x + 3)$ of a degree 2 and a degree 1 polynomial, and use the calculation to explain why Justin might be correct in general.

   c. Jason thinks he can generalize Justin's claim to the product of two polynomials. He claims that for $n \geq 2$, the $(n+1)^{\text{st}}$ differences of the product of a degree n polynomial and a linear polynomial are the same as the nth differences of the degree n polynomial. Use what you know about second and third differences (from Examples 2 and 3) and the polynomial $(5x^2 + x + 1)(2x + 3)$ to show that Jason's generalization is incorrect.

This page intentionally left blank

Lesson 2: The Multiplication of Polynomials

Classwork

Opening Exercise

Show that $28 \times 27 = (20 + 8)(20 + 7)$ using an area model. What do the numbers you placed inside the four rectangular regions you drew represent?

Example 1

Use the tabular method to multiply $(x + 8)(x + 7)$ and combine like terms.

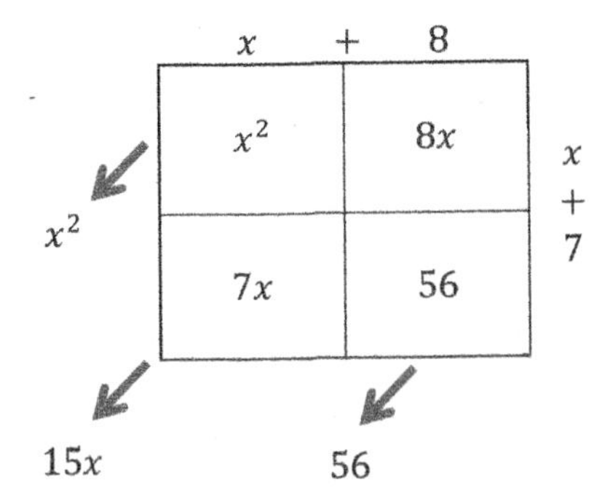

Exercises 1–2

1. Use the tabular method to multiply $(x^2 + 3x + 1)(x^2 - 5x + 2)$ and combine like terms.

2. Use the tabular method to multiply $(x^2 + 3x + 1)(x^2 - 2)$ and combine like terms.

Example 2

Multiply the polynomials $(x-1)(x^4+x^3+x^2+x+1)$ using a table. Generalize the pattern that emerges by writing down an identity for $(x-1)(x^n+x^{n-1}+\cdots+x^2+x+1)$ for n a positive integer.

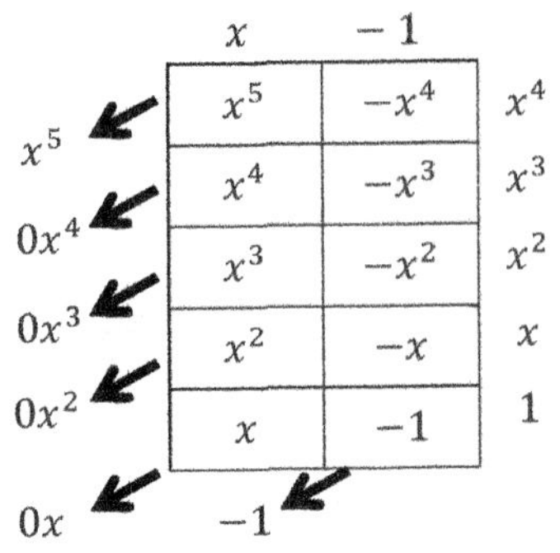

	x	-1	
x^5	x^5	$-x^4$	x^4
$0x^4$	x^4	$-x^3$	x^3
$0x^3$	x^3	$-x^2$	x^2
$0x^2$	x^2	$-x$	x
$0x$	x	-1	1
	-1		

Exercises 3–4

3. Multiply $(x-y)(x^3+x^2y+xy^2+y^3)$ using the distributive property and combine like terms. How is this calculation similar to Example 2?

4. Multiply $(x^2-y^2)(x^2+y^2)$ using the distributive property and combine like terms. Generalize the pattern that emerges to write down an identity for $(x^n-y^n)(x^n+y^n)$ for positive integers n.

Relevant Vocabulary

EQUIVALENT POLYNOMIAL EXPRESSIONS: Two polynomial expressions in one variable are *equivalent* if, whenever a number is substituted into all instances of the variable symbol in both expressions, the numerical expressions created are equal.

POLYNOMIAL IDENTITY: A *polynomial identity* is a statement that two polynomial expressions are equivalent. For example, $(x+3)^2 = x^2 + 6x + 9$ for any real number x is a polynomial identity.

COEFFICIENT OF A MONOMIAL: The *coefficient of a monomial* is the value of the numerical expression found by substituting the number 1 into all the variable symbols in the monomial. The coefficient of $3x^2$ is 3, and the coefficient of the monomial $(3xyz) \cdot 4$ is 12.

TERMS OF A POLYNOMIAL: When a polynomial is expressed as a monomial or a sum of monomials, each monomial in the sum is called a *term* of the polynomial.

LIKE TERMS OF A POLYNOMIAL: Two terms of a polynomial that have the same variable symbols each raised to the same power are called *like terms.*

STANDARD FORM OF A POLYNOMIAL IN ONE VARIABLE: A polynomial expression with one variable symbol, x, is in *standard form* if it is expressed as

$$a_n x^n + a_{n-1}x^{n-1} + \cdots + a_1 x + a_0,$$

where n is a non-negative integer, and $a_0, a_1, a_2 \ldots, a_n$ are constant coefficients with $a_n \neq 0$.

A polynomial expression in x that is in standard form is often just called a *polynomial in* x or a *polynomial.*

The *degree of the polynomial in standard form* is the highest degree of the terms in the polynomial, namely n. The term $a_n x^n$ is called the *leading term* and a_n (thought of as a specific number) is called the *leading coefficient.* The *constant term* is the value of the numerical expression found by substituting 0 into all the variable symbols of the polynomial, namely a_0.

Problem Set

1. Complete the following statements by filling in the blanks.
   a. $(a+b)(c+d+e) = ac + ad + ae + \_\_\_ + \_\_\_ + \_\_\_$
   b. $(r-s)^2 = (\_\_\_)^2 - (\_\_\_)rs + s^2$
   c. $(2x+3y)^2 = (2x)^2 + 2(2x)(3y) + (\_\_\_)^2$
   d. $(w-1)(1+w+w^2) = \_\_\_ - 1$
   e. $a^2 - 16 = (a + \_\_\_)(a - \_\_\_)$
   f. $(2x+5y)(2x-5y) = \_\_\_ - \_\_\_$
   g. $(2^{21}-1)(2^{21}+1) = \_\_\_ - 1$
   h. $[(x-y)-3][(x-y)+3] = (\_\_\_)^2 - 9$

2. Use the tabular method to multiply and combine like terms.
   a. $(x^2 - 4x + 4)(x + 3)$
   b. $(11 - 15x - 7x^2)(25 - 16x^2)$
   c. $(3m^3 + m^2 - 2m - 5)(m^2 - 5m - 6)$
   d. $(x^2 - 3x + 9)(x^2 + 3x + 9)$

3. Multiply and combine like terms to write as the sum or difference of monomials.
   a. $2a(5 + 4a)$
   b. $x^3(x + 6) + 9$
   c. $\frac{1}{8}(96z + 24z^2)$
   d. $2^{23}(2^{84} - 2^{81})$
   e. $(x - 4)(x + 5)$
   f. $(10w - 1)(10w + 1)$
   g. $(3z^2 - 8)(3z^2 + 8)$
   h. $(-5w - 3)w^2$
   i. $8y^{1000}(y^{12200} + 0.125y)$
   j. $(2r + 1)(2r^2 + 1)$
   k. $(t - 1)(t + 1)(t^2 + 1)$
   l. $(w - 1)(w^5 + w^4 + w^3 + w^2 + w + 1)$
   m. $(x + 2)(x + 2)(x + 2)$
   n. $n(n + 1)(n + 2)$
   o. $n(n + 1)(n + 2)(n + 3)$
   p. $n(n + 1)(n + 2)(n + 3)(n + 4)$
   q. $(x + 1)(x^3 - x^2 + x - 1)$
   r. $(x + 1)(x^5 - x^4 + x^3 - x^2 + x - 1)$
   s. $(x + 1)(x^7 - x^6 + x^5 - x^4 + x^3 - x^2 + x - 1)$
   t. $(m^3 - 2m + 1)(m^2 - m + 2)$

4. Polynomial expressions can be thought of as a generalization of place value.
   a. Multiply 214×112 using the standard paper-and-pencil algorithm.
   b. Multiply $(2x^2 + x + 4)(x^2 + x + 2)$ using the tabular method and combine like terms.
   c. Substitute $x = 10$ into your answer from part (b).
   d. Is the answer to part (c) equal to the answer from part (a)? Compare the digits you computed in the algorithm to the coefficients of the entries you computed in the table. How do the place-value units of the digits compare to the powers of the variables in the entries?

5. Jeremy says $(x - 9)(x^7 + x^6 + x^5 + x^4 + x^3 + x^2 + x + 1)$ must equal $x^7 + x^6 + x^5 + x^4 + x^3 + x^2 + x + 1$ because when $x = 10$, multiplying by $x - 9$ is the same as multiplying by 1.
   a. Multiply $(x - 9)(x^7 + x^6 + x^5 + x^4 + x^3 + x^2 + x + 1)$.
   b. Substitute $x = 10$ into your answer.
   c. Is the answer to part (b) the same as the value of $x^7 + x^6 + x^5 + x^4 + x^3 + x^2 + x + 1$ when $x = 10$?
   d. Was Jeremy right?

6. In the diagram, the side of the larger square is x units, and the side of the smaller square is y units. The area of the shaded region is $(x^2 - y^2)$ square units. Show how the shaded area might be cut and rearranged to illustrate that the area is $(x - y)(x + y)$ square units.

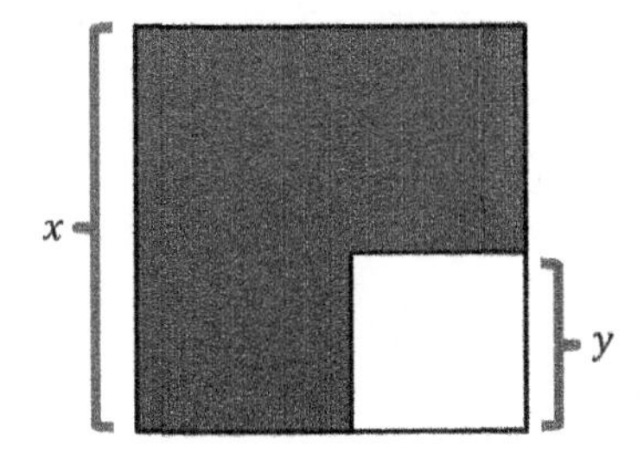

This page intentionally left blank

Lesson 3: The Division of Polynomials

Opening Exercise

a. Multiply these polynomials using the tabular method.

$$(2x+5)(x^2+5x+1)$$

b. How can you use the expression in part (a) to quickly multiply $25 \cdot 151$?

Exploratory Challenge

1. Does $\frac{2x^3+15x^2+27x+5}{2x+5} = (x^2+5x+1)$? Justify your answer.

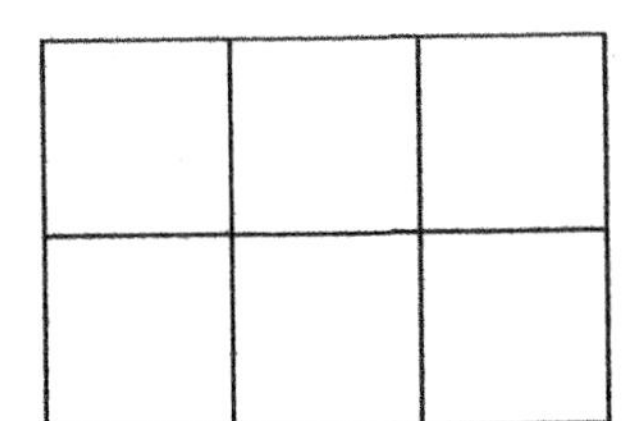

2. Describe the process you used to determine your answer to Exercise 1.

3. Reverse the tabular method of multiplication to find the quotient: $\frac{2x^2+x-10}{x-2}$.

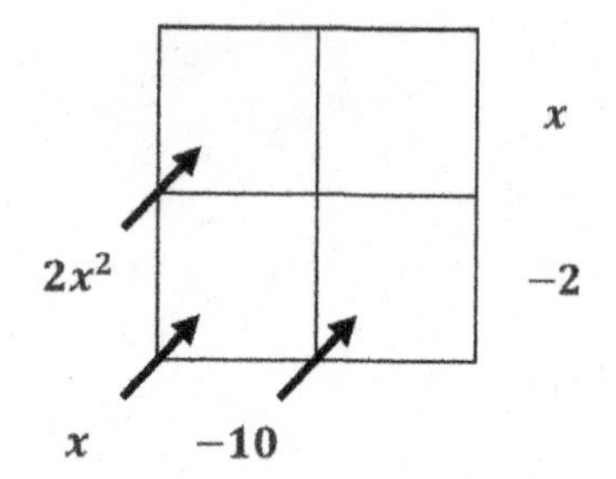

4. Test your conjectures. Create your own table and use the *reverse tabular method* to find the quotient.

$$\frac{x^4 + 4x^3 + 3x^2 + 4x + 2}{x^2 + 1}$$

5. Test your conjectures. Use the *reverse tabular method* to find the quotient.

$$\frac{3x^5 - 2x^4 + 6x^3 - 4x^2 - 24x + 16}{x^2 + 4}$$

6. What is the quotient of $\frac{x^5 - 1}{x - 1}$? What is the quotient of $\frac{x^6 - 1}{x - 1}$?

Problem Set

Use the reverse tabular method to solve these division problems.

1. $\dfrac{2x^3+x^2-16x+15}{2x-3}$

2. $\dfrac{3x^5+12x^4+11x^3+2x^2-4x-2}{3x^2-1}$

3. $\dfrac{x^3-4x^2+7x-28}{x^2+7}$

4. $\dfrac{x^4-2x^3-29x-12}{x^3+2x^2+8x+3}$

5. $\dfrac{6x^5+4x^4-6x^3+14x^2-8}{6x+4}$

6. $\dfrac{x^3-8}{x-2}$

7. $\dfrac{x^3+2x^2+2x+1}{x+1}$

8. $\dfrac{x^4+2x^3+2x^2+2x+1}{x+1}$

9. Use the results of Problems 7 and 8 to predict the quotient of $\dfrac{x^5+2x^4+2x^3+2x^2+2x+1}{x+1}$. Explain your prediction. Then check your prediction using the reverse tabular method.

10. Use the results of Problems 7–9 above to predict the quotient of $\dfrac{x^4-2x^3+2x^2-2x+1}{x-1}$. Explain your prediction. Then check your prediction using the reverse tabular method.

11. Make and test a conjecture about the quotient of $\dfrac{x^6+x^5+2x^4+2x^3+2x^2+x+1}{x^2+1}$. Explain your reasoning.

12. Consider the following quotients:

$$\frac{4x^2 + 8x + 3}{2x + 1} \text{ and } \frac{483}{21}$$

a. How are these expressions related?

b. Find each quotient.

c. Explain the connection between the quotients.

This page intentionally left blank

Lesson 4: Comparing Methods—Long Division, Again?

Opening Exercises

1. Use the reverse tabular method to determine the quotient $\frac{2x^3+11x^2+7x+10}{x+5}$.

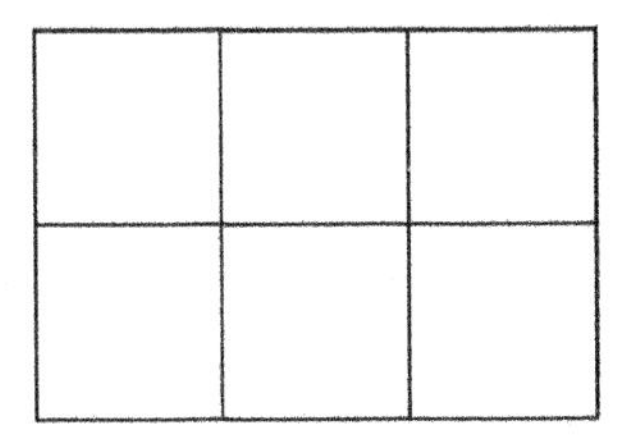

2. Use your work from Exercise 1 to write the polynomial $2x^3 + 11x^2 + 7x + 10$ in factored form, and then multiply the factors to check your work above.

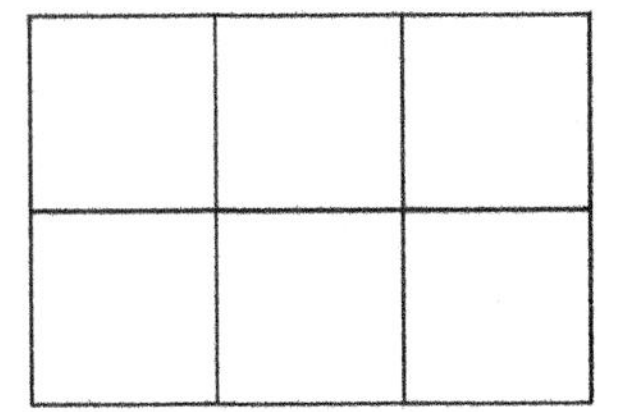

Example 1

If $x = 10$, then the division $1573 \div 13$ can be represented using polynomial division.

$$x+3\overline{\smash{)}x^3+5x^2+7x+3}$$

Example 2

Use the long division algorithm for polynomials to evaluate

$$\frac{2x^3-4x^2+2}{2x-2}.$$

Exercises 1–8

Use the long division algorithm to determine the quotient. For each problem, check your work by using the reverse tabular method.

1. $\frac{x^2+6x+9}{x+3}$

2. $\frac{7x^3-8x^2-13x+2}{7x-1}$

3. $\frac{x^3-27}{x-3}$

4. $\frac{2x^4+14x^3+x^2-21x-6}{2x^2-3}$

5. $\frac{5x^4-6x^2+1}{x^2-1}$

6. $\frac{x^6+4x^4-4x-1}{x^3-1}$

7. $\dfrac{2x^7+x^5-4x^3+14x^2-2x+7}{2x^2+1}$

8. $\dfrac{x^6-64}{x+2}$

Lesson Summary

The long division algorithm to divide polynomials is analogous to the long division algorithm for integers. The long division algorithm to divide polynomials produces the same results as the reverse tabular method.

Problem Set

Use the long division algorithm to determine the quotient in problems 1–5.

1. $\frac{2x^3-13x^2-x+3}{2x+1}$

2. $\frac{3x^3+4x^2+7x+22}{x+2}$

3. $\frac{x^4+6x^3-7x^2-24x+12}{x^2-4}$

4. $(12x^4 + 2x^3 + x - 3) \div (2x^2 + 1)$

5. $(2x^3 + 2x^2 + 2x) \div (x^2 + x + 1)$

6. Use long division to find the polynomial, p, that satisfies the equation below.

$$2x^4 - 3x^2 - 2 = (2x^2 + 1)(p(x))$$

7. Given $q(x) = 3x^3 - 4x^2 + 5x + k$.
   a. Determine the value of k so that $3x - 7$ is a factor of the polynomial q.
   b. What is the quotient when you divide the polynomial q by $3x - 7$?

8. In parts a–b and d–e, use long division to evaluate each quotient. Then, answer the remaining questions.
   a. $\frac{x^2-9}{x+3}$
   b. $\frac{x^4-81}{x+3}$
   c. Is $x+3$ a factor of x^3-27? Explain your answer using the long division algorithm.
   d. $\frac{x^3+27}{x+3}$
   e. $\frac{x^5+243}{x+3}$
   f. Is $x+3$ a factor of x^2+9? Explain your answer using the long division algorithm.
   g. For which positive integers n is $x+3$ a factor of x^n+3^n? Explain your reasoning.
   h. If n is a positive integer, is $x+3$ a factor of x^n-3^n? Explain your reasoning.

Lesson 5: Putting It All Together

Classwork

Exercises 1–15: Polynomial Pass

Perform the indicated operation to write each polynomial in standard form.

1. $(x^2 - 3)(x^2 + 3x - 1)$

2. $(5x^2 - 3x - 7) - (x^2 + 2x - 5)$

3. $\frac{x^3-8}{x-2}$

4. $(x + 1)(x - 2)(x + 3)$

5. $(x + 1) - (x - 2) - (x + 3)$

6. $(x + 2)(2x^2 - 5x + 7)$

7. $\frac{x^3-2x^2-65x+18}{x-9}$

8. $(x^2 - 3x + 2) - (2 - x + 2x^2)$

9. $(x^2 - 3x + 2)(2 - x + 2x^2)$

10. $\frac{x^3 - x^2 - 5x - 3}{x - 3}$

11. $(x^2 + 7x - 12)(x^2 - 9x + 1)$

12. $(2x^3 - 6x^2 - 7x - 2) + (x^3 + x^2 + 6x - 12)$

13. $(x^3 - 8)(x^2 - 4x + 4)$

14. $\frac{x^3 - 2x^2 - 5x + 6}{x + 2}$

15. $(x^3 + 2x^2 - 3x - 1) + (4 - x - x^3)$

Exercises 16–22

16. Review Exercises 1–15 and then select one exercise for each category and record the steps in the operation below as an example. Be sure to show all your work.

Addition Exercise	Multiplication Exercise
Subtraction Exercise	Division Exercise

For Exercises 17–20, rewrite each polynomial in standard form by applying the operations in the appropriate order.

17. $\frac{(x^2+5x+20)+(x^2+6x-6)}{x+2}$

18. $(x^2-4)(x+3)-(x^2+2x-5)$

19. $\frac{(x-3)^3}{x^2-6x+9}$

20. $(x+7)(2x-3)-(x^3-2x^2+x-2)\div(x-2)$

21. What would be the first and last terms of the polynomial if it was rewritten in standard form? Answer these quickly without performing all of the indicated operations.

    a. $(2x^3 - x^2 - 9x + 7) + (11x^2 - 6x^3 + 2x - 9)$

    b. $(x - 3)(2x + 3)(x - 1)$

    c. $(2x - 3)(3x + 5) - (x + 1)(2x^2 - 6x + 3)$

    d. $(x + 5)(3x - 1) - (x - 4)^2$

22. What would the first and last terms of the polynomial be if it was rewritten in standard form?

    a. $(n + 1)(n + 2)(n + 3) \vdots (n + 9)(n + 10)$

    b. $(x - 2)^{10}$

    c. $\frac{(x-2)^{10}}{(x-2)}$

    d. $\frac{n(n+1)(2n+1)}{6}$

Problem Set

For Problems 1–7, rewrite each expression as a polynomial in standard form.

1. $(3x-4)^3$

2. $(2x^2 - x^3 - 9x + 1) - (x^3 + 7x - 3x^2 + 1)$

3. $(x^2 - 5x + 2)(x - 3)$

4. $\frac{x^4 - x^3 - 6x^2 - 9x + 27}{x-3}$

5. $(x+3)(x-3) - (x+4)(x-4)$

6. $(x+3)^2 - (x+4)^2$

7. $\frac{x^2 - 5x + 6}{x-3} + \frac{x^3 - 1}{x-1}$

For Problems 8–9: Quick, what would be the first and last terms of the polynomial if it was written in standard form?

8. $2(x^2 - 5x + 4) - (x+3)(x+2)$

9. $\frac{(x-2)^5}{x-2}$

10. The profit a business earns by selling x items is given by the polynomial function

$$p(x) = x(160 - x) - (100x + 500).$$

What is the last term in the standard form of this polynomial? What does it mean in this situation?

11. Explain why these two quotients are different. Compute each one. What do they have in common? Why?

$$\frac{(x-2)^4}{x-2} \text{ and } \frac{x^4 - 16}{x-2}$$

12. What are the area and perimeter of the figure? Assume there is a right angle at each vertex.

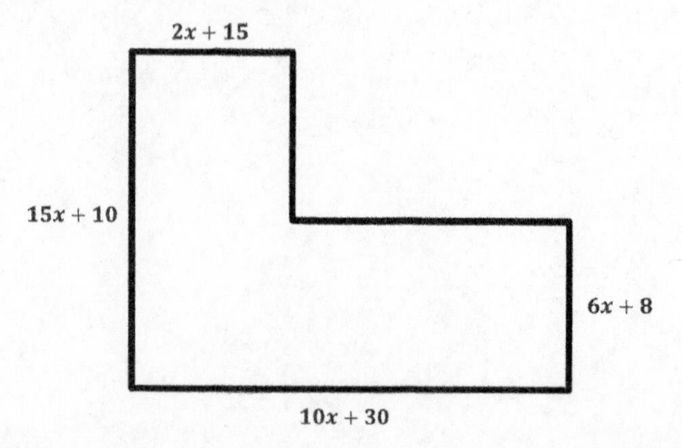

Lesson 6: Dividing by $x - a$ and by $x + a$

Classwork

Opening Exercise

Find the following quotients, and write the quotient in standard form.

a. $\frac{x^2-9}{x-3}$

b. $\frac{x^3-27}{x-3}$

c. $\frac{x^4-81}{x-3}$

Exercise 1

1. Use patterns to predict each quotient. Explain how you arrived at your prediction, and then test it by applying the reverse tabular method or long division.

   a. $\frac{x^2-144}{x-12}$

   b. $\frac{x^3-8}{x-2}$

   c. $\frac{x^3-125}{x-5}$

   d. $\frac{x^6-1}{x-1}$

Example 1

What is the quotient of $\frac{x^2-a^2}{x-a}$? Use the reverse tabular method or long division.

Exercises 2–4

2. Work with your group to find the following quotients.

   a. $\frac{x^3-a^3}{x-a}$

   b. $\frac{x^4-a^4}{x-a}$

3. Predict without performing division whether or not the divisor will divide into the dividend without a remainder for the following problems. If so, find the quotient. Then check your answer.

   a. $\frac{x^2-a^2}{x+a}$

   b. $\frac{x^3-a^3}{x+a}$

   c. $\frac{x^2+a^2}{x+a}$

   d. $\frac{x^3+a^3}{x+a}$

4.

a. Find the quotient $\frac{x^n-1}{x-1}$ for $n = 2, 3, 4,$ and 8

b. What patterns do you notice?

c. Use your work in part (a) to write an expression equivalent to $\frac{x^n-1}{x-1}$ for any integer $n > 1$.

Lesson Summary

Based on the work in this lesson, it can be concluded that the following statements are true for all real values of x and a:

$$x^2 - a^2 = (x - a)(x + a)$$
$$x^3 - a^3 = (x - a)(x^2 + ax + a^2)$$
$$x^3 + a^3 = (x + a)(x^2 - ax + a^2),$$

and it seems that the following statement is also an identity for all real values of x and a:

$x^n - 1 = (x - 1)(x^{n-1} + x^{n-2} + x^{n-3} + \cdots + x^1 + 1)$, for integers $n > 1$.

Problem Set

1. Compute each quotient.

   a. $\frac{x^2-625}{x-25}$

   b. $\frac{x^3+1}{x+1}$

   c. $\frac{x^3-\frac{1}{8}}{x-\frac{1}{2}}$

   d. $\frac{x^2-0.01}{x-0.1}$

2. In the next exercises, you can use the same identities you applied in the previous problem. Fill in the blanks in the problems below to help you get started. Check your work by using the reverse tabular method or long division to make sure you are applying the identities correctly.

   a. $\frac{16x^2-121}{4x-11} = \frac{(\_\_)^2-(\_\_)^2}{4x-11} = (\_\_) + 11$

   b. $\frac{25x^2-49}{5x+7} = \frac{(\_\_)^2-(\_\_)^2}{5x+7} = (\_\_) - (\_\_) =$ __________

   c. $\frac{8x^3-27}{2x-3} = \frac{(\_\_)^3-(\_\_)^3}{2x-3} = (\_\_)^2 + (\_\_)(\_\_) + (\_\_)^2 =$ __________

3. Show how the patterns and relationships learned in this lesson could be applied to solve the following arithmetic problems by filling in the blanks.

   a. $\frac{625-81}{16} = \frac{(\_)^2-(9)^2}{25-(\_)} = (\_) + (\_) = 34$

   b. $\frac{1000-27}{7} = \frac{(\_)^3-(\_)^3}{(\_)-3} = (\_)^2 + (10)(\_) + (\_)^2 = \_\_\_\_\_$

   c. $\frac{100-9}{7} = \frac{(\_)^2-(\_)^2}{(\_)-3} = \_\_\_\_\_$

   d. $\frac{1000+64}{14} = \frac{(\_)^3+(\_)^3}{(\_)+(\_)} = (\_)^2 - (\_)(\_) + (\_)^2 = \_\_\_\_\_$

4. Apply the identities from this lesson to compute each quotient. Check your work using the reverse tabular method or long division.

   a. $\frac{16x^2-9}{4x+3}$

   b. $\frac{81x^2-25}{18x-10}$

   c. $\frac{27x^3-8}{3x-2}$

5. Extend the patterns and relationships you learned in this lesson to compute the following quotients. Explain your reasoning, and then check your answer by using long division or the tabular method.

   a. $\frac{8+x^3}{2+x}$

   b. $\frac{x^4-y^4}{x-y}$

   c. $\frac{27x^3+8y^3}{3x+2y}$

   d. $\frac{x^7-y^7}{x-y}$

Lesson 7: Mental Math

Classwork

Opening Exercise

a. How are these two equations related?

$$\frac{x^2 - 1}{x + 1} = x - 1 \quad \text{and} \quad x^2 - 1 = (x + 1)(x - 1)$$

b. Explain the relationship between the polynomial identities $x^2 - 1 = (x + 1)(x - 1)$ and $x^2 - a^2 = (x - a)(x + a)$.

Exercises 1–3

1. Compute the following products using the identity $x^2 - a^2 = (x - a)(x + a)$. Show your steps.

   a. $6 \cdot 8$

   b. $11 \cdot 19$

c. $23 \cdot 17$

d. $34 \cdot 26$

2. Find two additional factors of $2^{100} - 1$.

3. Show that $8^3 - 1$ is divisible by 7.

Lesson Summary

Based on the work in this lesson, students can convert differences of squares into products (and vice versa) using

$$x^2 - a^2 = (x - a)(x + a).$$

If x, a, and n are integers with $(x - a) \neq \pm 1$ and $n > 1$, then numbers of the form $x^n - a^n$ are not prime because

$$x^n - a^n = (x - a)(x^{n-1} + ax^{n-2} + a^2x^{n-3} + \cdots + a^{n-2}x + a^{n-1}).$$

Problem Set

1. Using an appropriate polynomial identity, quickly compute the following products. Show each step. Be sure to state your values for x and a.
   a. $41 \cdot 19$
   b. $993 \cdot 1{,}007$
   c. $213 \cdot 187$
   d. $29 \cdot 51$
   e. $125 \cdot 75$

2. Give the general steps you take to determine x and a when asked to compute a product such as those in Problem 1.

3. Why is $17 \cdot 23$ easier to compute than $17 \cdot 22$?

4. Rewrite the following differences of squares as a product of two integers.
   a. $81 - 1$
   b. $400 - 121$

5. Quickly compute the following differences of squares.
   a. $64^2 - 14^2$
   b. $112^2 - 88^2$
   c. $785^2 - 215^2$

6. Is 323 prime? Use the fact that $18^2 = 324$ and an identity to support your answer.

7. The number $2^3 - 1$ is prime and so are $2^5 - 1$ and $2^7 - 1$. Does that mean $2^9 - 1$ is prime? Explain why or why not.

8. Show that $9{,}999{,}999{,}991$ is not prime without using a calculator or computer.

9. Show that 999,973 is not prime without using a calculator or computer.

10. Find a value of b so that the expression $b^n - 1$ is always divisible by 5 for any positive integer n. Explain why your value of b works for any positive integer n.

11. Find a value of b so that the expression $b^n - 1$ is always divisible by 7 for any positive integer n. Explain why your value of b works for any positive integer n.

12. Find a value of b so that the expression $b^n - 1$ is divisible by both 7 and 9 for any positive integer n. Explain why your value of b works for any positive integer n.

Lesson 8: The Power of Algebra—Finding Primes

Classwork

Opening Exercise: When is $2^n - 1$ prime and when is it composite?

Complete the table to investigate which numbers of the form $2^n - 1$ are prime and which are composite.

Exponent n	Expression $2^n - 1$	Value	Prime or Composite? Justify your answer if composite.
1			
2			
3			
4			
5			
6			
7			
8			
9			
10			
11			

What patterns do you notice in this table about which expressions are prime and which are composite?

Example 1: Proving a Conjecture

Conjecture: If m is a positive odd composite number, then $2^m - 1$ is a composite number.

Start with an identity: $x^n - 1 = (x-1)(x^{n-1} + x^{n-2} + \cdots x^1 + 1)$

In this case, $x = 2$, so the identity above becomes:

$$\begin{aligned} 2^m - 1 &= (2-1)(2^{m-1} + 2^{m-2} + \cdots + 2^1 + 1) \\ &= (2^{m-1} + 2^{m-2} + \cdots + 2^1 + 1), \end{aligned}$$

and it is not clear whether or not $2^m - 1$ is composite.

Rewrite the expression: Let $m = ab$ be a positive odd composite number. Then a and b must also be odd, or else the product ab would be even. The smallest such number m is 9, so we have $a \geq 3$ and $b \geq 3$.

Then we have

$$\begin{aligned} 2^m - 1 &= (2^a)^b - 1 \\ &= (2^a - 1)\underbrace{((2^a)^{b-1} + (2^a)^{b-2} + \cdots + (2^a)^1 + 1)}_{\text{Some number larger than 1}}. \end{aligned}$$

Since $a \geq 3$, we have $2^a \geq 8$; thus, $2^a - 1 \geq 7$. Since the other factor is also larger than 1, $2^m - 1$ is composite, and we have proven our conjecture.

Exercises 1–3

For Exercises 1–3, find a factor of each expression using the method discussed in Example 1.

1. $2^{15} - 1$

2. $2^{99} - 1$

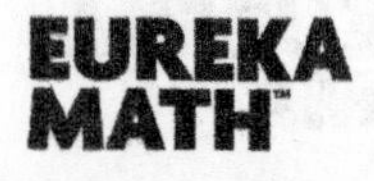

3. $2^{537} - 1$ (Hint: 537 is the product of two prime numbers that are both less than 50.)

Exercise 4: How quickly can a computer factor a very large number?

4. How long would it take a computer to factor some squares of very large prime numbers?

   The time in seconds required to factor an n-digit number of the form p^2, where p is a large prime, can roughly be approximated by $f(n) = 3.4 \times 10^{(n-13)/2}$. Some values of this function are listed in the table below.

p	p^2	Number of Digits	Time needed to factor the number (sec)
10,007	100,140,049	9	0.034
100,003	10,000,600,009	11	0.34
1,000,003	1,000,006,000,009	13	3.4
10,000,019	100,000,380,000,361	15	34
100,000,007	10,000,001,400,000,049	17	340
1000,000,007	1,000,000,014,000,000,049	19	3,400

   Use the function given above to determine how long it would take this computer to factor a number that contains 32 digits.

Problem Set

1. Factor $4^{12} - 1$ in two different ways using the identity $x^n - a^n = (x - a)(x^n + ax^{n-1} + a^2x^{n-2} + \cdots + a^n)$ and the difference of squares identity.

2. Factor $2^{12} + 1$ using the identity $x^n + a^n = (x + a)(x^n - ax^{n-1} + a^2x^{n-2} - \cdots + a^n)$ for odd numbers n.

3. Is $10{,}000{,}000{,}001$ prime? Explain your reasoning.

4. Explain why $2^n - 1$ is never prime if n is a composite number.

5. Fermat numbers are of the form $2^n + 1$ where n is a positive integer.
   a. Create a table of Fermat numbers for odd values of n up to 9.

n	$2^n + 1$
1	
3	
5	
7	
9	

   b. Explain why if n is odd, the Fermat number $2^n + 1$ will always be divisible by 3.

   c. Complete the table of values for even values of n up to 12.

n	$2^n + 1$
2	
4	
6	
8	
10	
12	

d. Show that if n can be written in the form $2k$ where k is odd, then $2^n + 1$ is divisible by 5.

e. Which even numbers are not divisible by an odd number? Make a conjecture about the only Fermat numbers that might be prime.

6. Complete this table to explore which numbers can be expressed as the difference of two perfect squares.

Number	Difference of Two Squares	Number	Difference of Two Squares
1	$1^2 - 0^2 = 1 - 0 = 1$	11	
2	Not possible	12	
3	$2^2 - 1^2 = 4 - 1 = 3$	13	
4	$2^2 - 0^2 = 4 - 0 = 4$	14	
5		15	
6		16	
7		17	
8		18	
9		19	
10		20	

a. For which odd numbers does it appear to be possible to write the number as the difference of two squares?

b. For which even numbers does it appear to be possible to write the number as the difference of two squares?

c. Suppose that n is an odd number that can be expressed as $n = a^2 - b^2$ for positive integers a and b. What do you notice about a and b?

d. Suppose that n is an even number that can be expressed as $n = a^2 - b^2$ for positive integers a and b. What do you notice about a and b?

7. Express the numbers from 21 to 30 as the difference of two squares, if possible.

8. Prove this conjecture: Every positive odd number m can be expressed as the difference of the squares of two consecutive numbers that sum to the original number m.

a. Let m be a positive odd number. Then for some integer n, $m = 2n + 1$. We will look at the consecutive integers n and $n + 1$. Show that $n + (n + 1) = m$.

b. What is the difference of squares of $n + 1$ and n?

c. What can you conclude from parts (a) and (b)?

9. Prove this conjecture: Every positive multiple of 4 can be expressed as the difference of squares of two numbers that differ by 2. Use the table below to organize your work for parts (a)–(c).
   a. Write each multiple of 4 in the table as a difference of squares.

n	$4n$	Difference of squares $a^2 - b^2$	a	b
1	4	$2^2 - 0^2$	2	0
2				
3				
4				
5				
n	$4n$	$(\quad)^2 - (\quad)^2$		

   b. What do you notice about the numbers a and b that are squared? How do they relate to the number n?
   c. Given a positive integer of the form $4n$, prove that there are integers a and b so that $4n = a^2 - b^2$ and that $a - b = 2$. (Hint: Refer to parts (a) and (b) for the relationship between n and a and b.)

10. The steps below prove that the only positive even numbers that can be written as a difference of square integers are the multiples of 4. That is, completing this exercise will prove that it is impossible to write a number of the form $4n - 2$ as a difference of square integers.
    a. Let m be a positive even integer that we can write as the difference of square integers $m = a^2 - b^2$. Then $m = (a + b)(a - b)$ for integers a and b. How do we know that either a and b are both even or a and b are both odd?
    b. Is $a + b$ even or odd? What about $a - b$? How do you know?
    c. Is 2 a factor of $a + b$? Is 2 a factor of $a - b$? Is 4 a factor of $(a + b)(a - b)$? Explain how you know.
    d. Is 4 a factor of any integer of the form $4n - 2$?
    e. What can you conclude from your work in parts a–d?

11. Explain why the prime number 17 can only be expressed as the difference of two squares in only one way, but the composite number 24 can be expressed as the difference of two squares in more than one way.

12. Explain why you cannot use the factors of 3 and 8 to rewrite 24 as the difference of two square integers.

Lesson 9: Radicals and Conjugates

Classwork

Opening Exercise

Which of these statements are true for all $a, b > 0$? Explain your conjecture.

i. $2(a + b) = 2a + 2b$

ii. $\frac{a+b}{2} = \frac{a}{2} + \frac{b}{2}$

iii. $\sqrt{a+b} = \sqrt{a} + \sqrt{b}$

Example 1

Express $\sqrt{50} - \sqrt{18} + \sqrt{8}$ in simplest radical form and combine like terms.

Exercises 1–5

1. $\sqrt{\frac{1}{4}} + \sqrt{\frac{9}{4}} - \sqrt{45}$

2. $\sqrt{2}\left(\sqrt{3} - \sqrt{2}\right)$

3. $\sqrt{\frac{3}{8}}$

4. $\sqrt[3]{\frac{5}{32}}$

5. $\sqrt[3]{16x^5}$

Example 2

Multiply and combine like terms. Then explain what you notice about the two different results.

$(\sqrt{3}+\sqrt{2})(\sqrt{3}+\sqrt{2})$

$(\sqrt{3}+\sqrt{2})(\sqrt{3}-\sqrt{2})$

Exercise 6

6. Find the product of the conjugate radicals.

   $(\sqrt{5}+\sqrt{3})(\sqrt{5}-\sqrt{3})$

   $(7+\sqrt{2})(7-\sqrt{2})$

   $(\sqrt{5}+2)(\sqrt{5}-2)$

Example 3

Write $\frac{\sqrt{3}}{5-2\sqrt{3}}$ in simplest radical form.

Lesson Summary

- For real numbers $a \geq 0$ and $b \geq 0$, where $b \neq 0$ when b is a denominator,
$$\sqrt{ab} = \sqrt{a} \cdot \sqrt{b} \text{ and } \frac{\sqrt{a}}{\sqrt{b}} = \frac{\sqrt{a}}{\sqrt{b}}.$$
- For real numbers $a \geq 0$ and $b \geq 0$, where $b \neq 0$ when b is a denominator,
$$\sqrt[3]{ab} = \sqrt[3]{a} \cdot \sqrt[3]{b} \text{ and } \sqrt[3]{\frac{a}{b}} = \frac{\sqrt[3]{a}}{\sqrt[3]{b}}.$$
- Two binomials of the form $\sqrt{a} + \sqrt{b}$ and $\sqrt{a} - \sqrt{b}$ are called conjugate radicals:
  $\sqrt{a} + \sqrt{b}$ is the conjugate of $\sqrt{a} - \sqrt{b}$, and
  $\sqrt{a} - \sqrt{b}$ is the conjugate of $\sqrt{a} + \sqrt{b}$.
  For example, the conjugate of $2 - \sqrt{3}$ is $2 + \sqrt{3}$.
- To rewrite an expression with a denominator of the form $\sqrt{a} + \sqrt{b}$ in simplest radical form, multiply the numerator and denominator by the conjugate $\sqrt{a} - \sqrt{b}$ and combine like terms.

Problem Set

1. Express each of the following as a rational number or in simplest radical form. Assume that the symbols a, b, and x represent positive numbers.
   a. $\sqrt{36}$
   b. $\sqrt{72}$
   c. $\sqrt{18}$
   d. $\sqrt{9x^3}$
   e. $\sqrt{27x^2}$
   f. $\sqrt[3]{16}$
   g. $\sqrt[3]{24a}$
   h. $\sqrt{9a^2 + 9b^2}$

2. Express each of the following in simplest radical form, combining terms where possible.
   a. $\sqrt{25} + \sqrt{45} - \sqrt{20}$
   b. $3\sqrt{3} - \sqrt{\frac{3}{4}} + \sqrt{\frac{1}{3}}$
   c. $\sqrt[3]{54} - \sqrt[3]{8} + 7\sqrt[3]{\frac{1}{4}}$
   d. $\sqrt[3]{\frac{5}{8}} + \sqrt[3]{40} - \sqrt[3]{\frac{8}{9}}$

3. Evaluate $\sqrt{x^2 - y^2}$ when $x = 33$ and $y = 15$.

4. Evaluate $\sqrt{x^2 + y^2}$ when $x = 20$ and $y = 10$.

5. Express each of the following as a rational expression or in simplest radical form. Assume that the symbols x and y represent positive numbers.
   a. $\sqrt{3}(\sqrt{7} - \sqrt{3})$
   b. $(3 + \sqrt{2})^2$
   c. $(2 + \sqrt{3})(2 - \sqrt{3})$
   d. $(2 + 2\sqrt{5})(2 - 2\sqrt{5})$
   e. $(\sqrt{7} - 3)(\sqrt{7} + 3)$
   f. $(3\sqrt{2} + \sqrt{7})(3\sqrt{2} - \sqrt{7})$
   g. $(x - \sqrt{3})(x + \sqrt{3})$
   h. $(2x\sqrt{2} + y)(2x\sqrt{2} - y)$

6. Simplify each of the following quotients as far as possible.
   a. $(\sqrt{21} - \sqrt{3}) \div \sqrt{3}$
   b. $(\sqrt{5} + 4) \div (\sqrt{5} + 1)$
   c. $(3 - \sqrt{2}) \div (3\sqrt{2} - 5)$
   d. $(2\sqrt{5} - \sqrt{3}) \div (3\sqrt{5} - 4\sqrt{2})$

7. If $x = 2 + \sqrt{3}$, show that $x + \frac{1}{x}$ has a rational value.

8. Evaluate $5x^2 - 10x$ when the value of x is $\frac{2-\sqrt{5}}{2}$.

9. Write the factors of $a^4 - b^4$. Express $(\sqrt{3} + \sqrt{2})^4 - (\sqrt{3} - \sqrt{2})^4$ in a simpler form.

10. The converse of the Pythagorean theorem is also a theorem: If the square of one side of a triangle is equal to the sum of the squares of the other two sides, then the triangle is a right triangle.

    Use the converse of the Pythagorean theorem to show that for $A, B, C > 0$, if $A + B = C$, then $\sqrt{A} + \sqrt{B} > \sqrt{C}$, so that $\sqrt{A} + \sqrt{B} > \sqrt{A + B}$.

This page intentionally left blank

Lesson 10: The Power of Algebra—Finding Pythagorean Triples

Classwork

Opening Exercise

Sam and Jill decide to explore a city. Both begin their walk from the same starting point.

- Sam walks 1 block north, 1 block east, 3 blocks north, and 3 blocks west.
- Jill walks 4 blocks south, 1 block west, 1 block north, and 4 blocks east.

If all city blocks are the same length, who is the farthest distance from the starting point?

Example 1

Prove that if $x > 1$, then a triangle with side lengths $x^2 - 1$, $2x$, and $x^2 + 1$ is a right triangle.

Example 2

Next we describe an easy way to find Pythagorean triples using the expressions from Example 1. Look at the multiplication table below for $\{1, 2, \ldots, 9\}$. Notice that the square numbers $\{1, 4, 9, \ldots, 81\}$ lie on the diagonal of this table.

a. What value of x is used to generate the Pythagorean triple $(15,8,17)$ by the formula $(x^2 - 1,\ 2x,\ x^2 + 1)$? How do the numbers $(1, 4, 4, 16)$ at the corners of the shaded square in the table relate to the values 15, 8, and 17?

×	1	2	3	4	5	6	7	8	9	10
1	**1**	2	3	**4**	5	6	7	8	9	10
2	2	4	6	8	10	12	14	16	18	20
3	3	6	9	12	15	18	21	24	27	30
4	**4**	8	12	**16**	20	24	28	32	36	40
5	5	10	15	20	25	30	35	40	45	50
6	6	12	18	24	30	36	42	48	54	60
7	7	14	21	28	35	42	49	56	63	70
8	8	16	24	32	40	48	56	64	72	80
9	9	18	27	36	45	54	63	72	81	90
10	10	20	30	40	50	60	70	80	90	100

b. Now you try one. Form a square on the multiplication table below whose left-top corner is the 1 (as in the example above) and whose bottom-right corner is a square number. Use the sums or differences of the numbers at the vertices of your square to form a Pythagorean triple. Check that the triple you generate is a Pythagorean triple.

×	1	2	3	4	5	6	7	8	9	10
1	1	2	3	4	5	6	7	8	9	10
2	2	4	6	8	10	12	14	16	18	20
3	3	6	9	12	15	18	21	24	27	30
4	4	8	12	16	20	24	28	32	36	40
5	5	10	15	20	25	30	35	40	45	50
6	6	12	18	24	30	36	42	48	54	60
7	7	14	21	28	35	42	49	56	63	70
8	8	16	24	32	40	48	56	64	72	80
9	9	18	27	36	45	54	63	72	81	90
10	10	20	30	40	50	60	70	80	90	100

Let's generalize this square to any square in the multiplication table where two opposite vertices of the square are square numbers.

c. How can you use the sums or differences of the numbers at the vertices of the shaded square to get a triple $(16, 30, 34)$? Is this a Pythagorean triple?

×	1	2	3	4	5	6	7	8	9	10
1	1	2	3	4	5	6	7	8	9	10
2	2	4	6	8	10	12	14	16	18	20
3	3	6	**9**	12	**15**	18	21	24	27	30
4	4	8	12	16	20	24	28	32	36	40
5	5	10	**15**	20	**25**	30	35	40	45	50
6	6	12	18	24	30	36	42	48	54	60
7	7	14	21	28	35	42	49	56	63	70
8	8	16	24	32	40	48	56	64	72	80
9	9	18	27	36	45	54	63	72	81	90
10	10	20	30	40	50	60	70	80	90	100

d. Using x instead of 5 and y instead of 3 in your calculations in part (c), write down a formula for generating Pythagorean triples in terms of x and y.

Relevant Facts and Vocabulary

PYTHAGOREAN THEOREM: If a right triangle has legs of length a and b units and hypotenuse of length c units, then $a^2 + b^2 = c^2$.

CONVERSE TO THE PYTHAGOREAN THEOREM: If the lengths a, b, c of the sides of a triangle are related by $a^2 + b^2 = c^2$, then the angle opposite the side of length c is a right angle.

PYTHAGOREAN TRIPLE: A *Pythagorean triple* is a triplet of positive integers (a, b, c) such that $a^2 + b^2 = c^2$. The triplet $(3, 4, 5)$ is a Pythagorean triple but $(1, 1, \sqrt{2})$ is not, even though the numbers are side lengths of an isosceles right triangle.

Problem Set

1. Rewrite each expression as a sum or difference of terms.
   a. $(x-3)(x+3)$
   b. $(x^2-3)(x^2+3)$
   c. $(x^{15}+3)(x^{15}-3)$
   d. $(x-3)(x^2+9)(x+3)$
   e. $(x^2+y^2)(x^2-y^2)$
   f. $(x^2+y^2)^2$
   g. $(x-y)^2(x+y)^2$
   h. $(x-y)^2(x^2+y^2)^2(x+y)^2$

2. Tasha used a clever method to expand $(a+b+c)(a+b-c)$. She grouped the addends together like this $[(a+b)+c][(a+b)-c]$ and then expanded them to get the difference of two squares:
$$(a+b+c)(a+b-c)=[(a+b)+c][(a+b)-c]=(a+b)^2-c^2=a^2+2ab+b^2-c^2.$$
   a. Is Tasha's method correct? Explain why or why not.
   b. Use a version of her method to find $(a+b+c)(a-b-c)$.
   c. Use a version of her method to find $(a+b-c)(a-b+c)$.

3. Use the difference of two squares identity to factor each of the following expressions.
   a. x^2-81
   b. $(3x+y)^2-(2y)^2$
   c. $4-(x-1)^2$
   d. $(x+2)^2-(y+2)^2$

4. Show that the expression $(x+y)(x-y)-6x+9$ may be written as the difference of two squares, and then factor the expression.

5. Show that $(x+y)^2-(x-y)^2=4xy$ for all real numbers x and y.

6. Prove that a triangle with side lengths x^2-y^2, $2xy$, and x^2+y^2 with $x>y>0$ is a right triangle.

7. Complete the table below to find Pythagorean triples (the first row is done for you).

x	y	x^2-y^2	$2xy$	x^2+y^2	Check: Is it a Pythagorean Triple?
2	1	3	4	5	Yes: $3^2+4^2=25=5^2$
3	1				
3	2				
4	1				
4	2				
4	3				
5	1				

8. Answer the following parts about the triple $(9,12,15)$.

   a. Show that $(9, 12, 15)$ is a Pythagorean triple.

   b. Prove that neither $(9, 12, 15)$ nor $(12,9,15)$ can be found by choosing a pair of integers x and y with $x > y$ and computing $(x^2 - y^2, 2xy, x^2 + y^2)$.
   (Hint: What are the possible values of x and y if $2xy = 12$? What about if $2xy = 9$?)

   c. Wouldn't it be nice if all Pythagorean triples were generated by $(x^2 - y^2, 2xy,\ x^2 + y^2)$? Research Pythagorean triples on the Internet to discover what is known to be true about generating all Pythagorean triples using this formula.

9. Follow the steps below to prove the identity $\left(a^2 + b^2\right)(x^2 + y^2) = (ax - by)^2 + (bx + ay)^2$.

   a. Multiply $\left(a^2 + b^2\right)(x^2 + y^2)$.

   b. Square both binomials in $(ax - by)^2 + (bx + ay)^2$ and collect like terms.

   c. Use your answers from part (a) and part (b) to prove the identity.

10. Many U.S. presidents took great delight in studying mathematics. For example, President James Garfield, while still a congressman, came up with a proof of the Pythagorean theorem based upon the ideas presented below.

    In the diagram, two congruent right triangles with side lengths a, b, and hypotenuse c, are used to form a trapezoid $PQRS$ composed of three triangles.

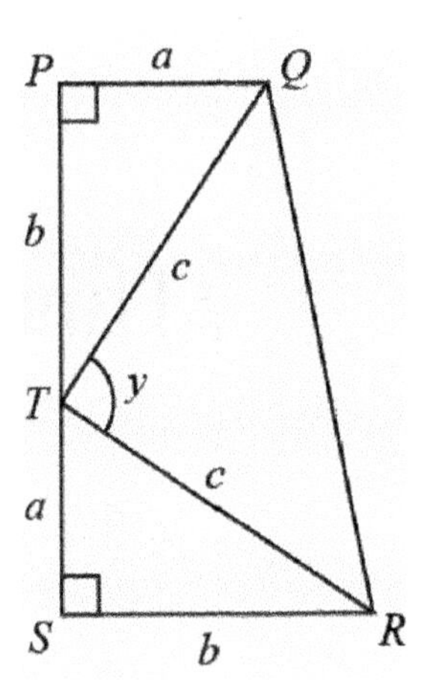

   a. Explain why $\angle QTR$ is a right angle.

   b. What are the areas of $\triangle\ STR$, $\triangle\ PTQ$, and $\triangle\ QTR$ in terms of a, b, and c?

   c. Using the formula for the area of a trapezoid, what is the total area of trapezoid $PQRS$ in terms of a and b?

   d. Set the sum of the areas of the three triangles from part (b) equal to the area of the trapezoid you found in part (c), and simplify the equation to derive a relationship between a, b, and c. Conclude that a right triangle with legs of length a and b and hypotenuse of length c must satisfy the relationship $a^2 + b^2 = c^2$.

This page intentionally left blank

Lesson 11: The Special Role of Zero in Factoring

Classwork

Opening Exercise

Find all solutions to the equation $(x^2 + 5x + 6)(x^2 - 3x - 4) = 0$.

Exercise 1

1. Find the solutions of $(x^2 - 9)(x^2 - 16) = 0$.

Example 1

Suppose we know that the polynomial equation $4x^3 - 12x^2 + 3x + 5 = 0$ has three real solutions and that one of the factors of $4x^3 - 12x^2 + 3x + 5$ is $(x - 1)$. How can we find all three solutions to the given equation?

Exercises 2–5

2. Find the zeros of the following polynomial functions, with their multiplicities.

   a. $f(x) = (x+1)(x-1)(x^2+1)$

   b. $g(x) = (x-4)^3(x-2)^8$

   c. $h(x) = (2x-3)^5$

   d. $k(x) = (3x+4)^{100}(x-17)^4$

3. Find a polynomial function that has the following zeros and multiplicities. What is the degree of your polynomial?

Zero	Multiplicity
2	3
−4	1
6	6
−8	10

4. Is there more than one polynomial function that has the same zeros and multiplicities as the one you found in Exercise 3?

5. Can you find a rule that relates the multiplicities of the zeros to the degree of the polynomial function?

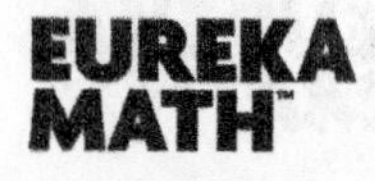

Relevant Vocabulary Terms

In the definitions below, the symbol $\mathbb{R}$ stands for the set of real numbers.

FUNCTION: A *function* is a correspondence between two sets, X and Y, in which each element of X is assigned to one and only one element of Y.

The set X in the definition above is called the *domain of the function.* The *range (or image)* of the function is the subset of Y, denoted $f(X)$, that is defined by the following property: y is an element of $f(X)$ if and only if there is an x in X such that $f(x) = y$.

If $f(x) = x^2$ where x can be any real number, then the domain is all real numbers (denoted $\mathbb{R}$), and the range is the set of nonnegative real numbers.

POLYNOMIAL FUNCTION: Given a polynomial expression in one variable, a *polynomial function in one variable* is a function $f: \mathbb{R} \to \mathbb{R}$ such that for each real number x in the domain, $f(x)$ is the value found by substituting the number x into all instances of the variable symbol in the polynomial expression and evaluating.

It can be shown that if a function $f: \mathbb{R} \to \mathbb{R}$ is a polynomial function, then there is some non-negative integer n and collection of real numbers $a_0, a_1, a_2, \dots, a_n$ with $a_n \neq 0$ such that the function satisfies the equation

$$f(x) = a_n x^n + a_{n-1}x^{n-1} + \cdots + a_1 x + a_0,$$

for every real number x in the domain, which is called the *standard form of the polynomial function.* The function $f(x) = 3x^3 + 4x^2 + 4x + 7$, where x can be any real number, is an example of a function written in standard form.

DEGREE OF A POLYNOMIAL FUNCTION: The *degree of a polynomial function* is the degree of the polynomial expression used to define the polynomial function.

The degree of $f(x) = 8x^3 + 4x^2 + 7x + 6$ is 3, but the degree of $g(x) = (x+1)^2 - (x-1)^2$ is 1 because when g is put into standard form, it is $g(x) = 4x$.

CONSTANT FUNCTION: A *constant function* is a polynomial function of degree 0. A constant function is of the form $f(x) = c$, for a constant c.

LINEAR FUNCTION: A *linear function* is a polynomial function of degree 1. A linear function is of the form $f(x) = ax + b$, for constants a and b with $a \neq 0$.

QUADRATIC FUNCTION: A *quadratic function* is a polynomial function of degree 2. A quadratic function is in *standard form* if it is written in the form $f(x) = ax^2 + bx + c$, for constants a, b, c with $a \neq 0$ and any real number x.

CUBIC FUNCTION: A *cubic function* is a polynomial function of degree 3. A cubic function is of the form $f(x) = ax^3 + bx^2 + cx + d$, for constants a, b, c, d with $a \neq 0$.

ZEROS OR ROOTS OF A FUNCTION: A *zero* (or *root*) of a function $f: \mathbb{R} \to \mathbb{R}$ is a number x of the domain such that $f(x) = 0$. A zero of a function is an element in the solution set of the equation $f(x) = 0$.

Lesson Summary

Given any two polynomial functions p and q, the solution set of the equation $p(x)q(x) = 0$ can be quickly found by solving the two equations $p(x) = 0$ and $q(x) = 0$ and combining the solutions into one set.

The number a is a zero of a polynomial function p with multiplicity m if the factored form of p contains $(x - a)^m$.

Problem Set

For Problems 1–4, find all solutions to the given equations.

1. $(x - 3)(x + 2) = 0$

2. $(x - 5)(x + 2)(x + 3) = 0$

3. $(2x - 4)(x + 5) = 0$

4. $(2x - 2)(3x + 1)(x - 1) = 0$

5. Find four solutions to the equation $(x^2 - 9)(x^4 - 16) = 0$.

6. Find the zeros with multiplicity for the function $p(x) = (x^3 - 8)(x^5 - 4x^3)$.

7. Find two different polynomial functions that have zeros at 1, 3, and 5 of multiplicity 1.

8. Find two different polynomial functions that have a zero at 2 of multiplicity 5 and a zero at -4 of multiplicity 3.

9. Find three solutions to the equation $(x^2 - 9)(x^3 - 8) = 0$.

10. Find two solutions to the equation $(x^3 - 64)(x^5 - 1) = 0$.

11. If p, q, r, s are nonzero numbers, find the solutions to the equation $(px + q)(rx + s) = 0$ in terms of p, q, r, s.

Use the identity $a^2 - b^2 = (a - b)(a + b)$ to solve the equations given in Problems 12–13.

12. $(3x - 2)^2 = (5x + 1)^2$

13. $(x + 7)^2 = (2x + 4)^2$

14. Consider the polynomial function $P(x) = x^3 + 2x^2 + 2x - 5$.
    a. Divide P by the divisor $(x - 1)$ and rewrite in the form $P(x) = (\text{divisor})(\text{quotient}) + \text{remainder}$.
    b. Evaluate $P(1)$.

15. Consider the polynomial function $Q(x) = x^6 - 3x^5 + 4x^3 - 12x^2 + x - 3$.
    a. Divide Q by the divisor $(x - 3)$ and rewrite in the form $Q(x) = (\text{divisor})(\text{quotient}) + \text{remainder}$.
    b. Evaluate $Q(3)$.

16. Consider the polynomial function $R(x) = x^4 + 2x^3 - 2x^2 - 3x + 2$.
    a. Divide R by the divisor $(x + 2)$ and rewrite in the form $R(x) = (\text{divisor})(\text{quotient}) + \text{remainder}$.
    b. Evaluate $R(-2)$.

17. Consider the polynomial function $S(x) = x^7 + x^6 - x^5 - x^4 + x^3 + x^2 - x - 1$.
    a. Divide S by the divisor $(x + 1)$ and rewrite in the form $S(x) = (\text{divisor})(\text{quotient}) + \text{remainder}$.
    b. Evaluate $S(-1)$.

18. Make a conjecture based on the results of Problems 14—17.

This page intentionally left blank

Lesson 12: Overcoming Obstacles in Factoring

Classwork

Example 1

Find all real solutions to the equation $(x^2 - 6x + 3)(2x^2 - 4x - 7) = 0$.

Exercise 1

Factor and find all real solutions to the equation $(x^2 - 2x - 4)(3x^2 + 8x - 3) = 0$.

Example 2

Find all solutions to $x^3 + 3x^2 - 9x - 27 = 0$ by factoring the equation.

Exercise 2

Find all real solutions to $x^3 - 5x^2 - 4x + 20 = 0$.

Exercise 3

Find all real solutions to $x^3 - 8x^2 - 2x + 16 = 0$.

Lesson Summary

In this lesson, we learned some techniques to use when faced with factoring polynomials and solving polynomial equations.

- If a fourth-degree polynomial can be factored into two quadratic expressions, then each quadratic expression might be factorable either using the quadratic formula or by completing the square.
- Some third-degree polynomials can be factored using the technique of factoring by grouping.

Problem Set

1. Solve each of the following equations by completing the square.
   a. $x^2 - 6x + 2 = 0$
   b. $x^2 - 4x = -1$
   c. $x^2 + x - \frac{3}{4} = 0$
   d. $3x^2 - 9x = -6$
   e. $(2x^2 - 5x + 2)(3x^2 - 4x + 1) = 0$
   f. $x^4 - 4x^2 + 2 = 0$

2. Solve each of the following equations using the quadratic formula.
   a. $x^2 - 5x - 3 = 0$
   b. $(6x^2 - 7x + 2)(x^2 - 5x + 5) = 0$
   c. $(3x^2 - 13x + 14)(x^2 - 4x + 1) = 0$

3. Not all of the expressions in the equations below can be factored using the techniques discussed so far in this course. First, determine if the expression can be factored with real coefficients. If so, factor the expression, and find all real solutions to the equation.
   a. $x^2 - 5x - 24 = 0$
   b. $3x^2 + 5x - 2 = 0$
   c. $x^2 + 2x + 4 = 0$
   d. $x^3 + 3x^2 - 2x + 6 = 0$
   e. $x^3 + 3x^2 + 2x + 6 = 0$
   f. $2x^3 + x^2 - 6x - 3 = 0$
   g. $8x^3 - 12x^2 + 2x - 3 = 0$
   h. $6x^3 + 8x^2 + 15x + 20 = 0$
   i. $4x^3 + 2x^2 - 36x - 18 = 0$
   j. $x^2 - \frac{1}{2}x - \frac{15}{2} = 0$

4. Solve the following equations by bringing all terms to one side of the equation and factoring out the greatest common factor.
   a. $(x-2)(x-1) = (x-2)(x+1)$
   b. $(2x+3)(x-4) = (2x+3)(x+5)$
   c. $(x-1)(2x+3) = (x-1)(x+2)$
   d. $(x^2+1)(3x-7) = (x^2+1)(3x+2)$
   e. $(x+3)(2x^2+7) = (x+3)(x^2+8)$

5. Consider the expression x^4+1. Since x^2+1 does not factor with real number coefficients, we might expect that x^4+1 also does not factor with real number coefficients. In this exercise, we investigate the possibility of factoring x^4+1.
   a. Simplify the expression $(x^2+1)^2 - 2x^2$.
   b. Factor $(x^2+1)^2 - 2x^2$ as a difference of squares.
   c. Is it possible to factor x^4+1 with real number coefficients? Explain

Lesson 13: Mastering Factoring

Classwork

Opening Exercises

Factor each of the following expressions. What similarities do you notice between the examples in the left column and those on the right?

a. $x^2 - 1$

b. $9x^2 - 1$

c. $x^2 + 8x + 15$

d. $4x^2 + 16x + 15$

e. $x^2 - y^2$

f. $x^4 - y^4$

Example 1

Write $9 - 16x^4$ as the product of two factors.

Example 2

Factor $4x^2y^4 - 25x^4z^6$.

Exercise 1

1. Factor the following expressions:
   a. $4x^2 + 4x - 63$

   b. $12y^2 - 24y - 15$

Exercises 2–4

Factor each of the following, and show that the factored form is equivalent to the original expression.

2. $a^3 + 27$

3. $x^3 - 64$

4. $2x^3 + 128$

Lesson Summary

In this lesson we learned additional strategies for factoring polynomials.

- The difference of squares identity $a^2 - b^2 = (a - b)(a + b)$ can be used to factor more advanced binomials.
- Trinomials can often be factored by looking for structure and then applying our previous factoring methods.
- Sums and differences of cubes can be factored by the formulas

$$x^3 + a^3 = (x + a)(x^2 - ax - a^2)$$
$$x^3 - a^3 = (x - a)(x^2 + ax + a^2).$$

Problem Set

1. If possible, factor the following expressions using the techniques discussed in this lesson.
   a. $25x^2 - 25x - 14$
   b. $9x^2y^2 - 18xy + 8$
   c. $45y^2 + 15y - 10$
   d. $y^6 - y^3 - 6$
   e. $x^3 - 125$
   f. $2x^4 - 16x$
   g. $9x^2 - 25y^4z^6$
   h. $36x^6y^4z^2 - 25x^2z^{10}$
   i. $4x^2 + 9$
   j. $x^4 - 36$
   k. $1 + 27x^9$
   l. $x^3y^6 + 8z^3$

2. Consider the polynomial expression $y^4 + 4y^2 + 16$.
   a. Is $y^4 + 4y^2 + 16$ factorable using the methods we have seen so far?
   b. Factor $y^6 - 64$ first as a difference of cubes, and then factor completely: $(y^2)^3 - 4^3$.
   c. Factor $y^6 - 64$ first as a difference of squares, and then factor completely: $(y^3)^2 - 8^2$.
   d. Explain how your answers to parts (b) and (c) provide a factorization of $y^4 + 4y^2 + 16$.
   e. If a polynomial can be factored as either a difference of squares or a difference of cubes, which formula should you apply first, and why?

3. Create expressions that have a structure that allows them to be factored using the specified identity. Be creative, and produce challenging problems!
   a. Difference of squares
   b. Difference of cubes
   c. Sum of cubes

Lesson 14: Graphing Factored Polynomials

Classwork

Opening Exercise

An engineer is designing a roller coaster for younger children and has tried some functions to model the height of the roller coaster during the first 300 yards. She came up with the following function to describe what she believes would make a fun start to the ride:

$$H(x) = -3x^4 + 21x^3 - 48x^2 + 36x,$$

where $H(x)$ is the height of the roller coaster (in yards) when the roller coaster is $100x$ yards from the beginning of the ride. Answer the following questions to help determine at which distances from the beginning of the ride the roller coaster is at its lowest height.

a. Does this function describe a roller coaster that would be fun to ride? Explain.

b. Can you see any obvious x-values from the equation where the roller coaster is at height 0?

c. Using a graphing utility, graph the function H on the interval $0 \leq x \leq 3$, and identify when the roller coaster is 0 yards off the ground.

d. What do the x-values you found in part (c) mean in terms of distance from the beginning of the ride?

e. Why do roller coasters always start with the largest hill first?

f. Verify your answers to part (c) by factoring the polynomial function H.

g. How do you think the engineer came up with the function for this model?

h. What is wrong with this roller coaster model at distance 0 yards and 300 yards? Why might this not initially bother the engineer when she is first designing the track?

Example 1

Graph each of the following polynomial functions. What are the function's zeros (counting multiplicities)? What are the solutions to $f(x) = 0$? What are the x-intercepts to the graph of the function? How does the degree of the polynomial function compare to the x-intercepts of the graph of the function?

a. $f(x) = x(x-1)(x+1)$

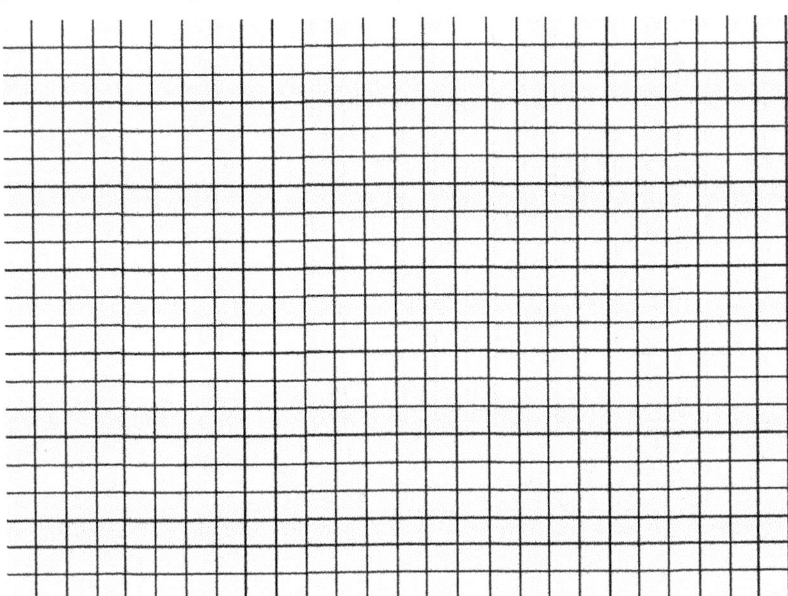

b. $f(x) = (x+3)(x+3)(x+3)(x+3)$

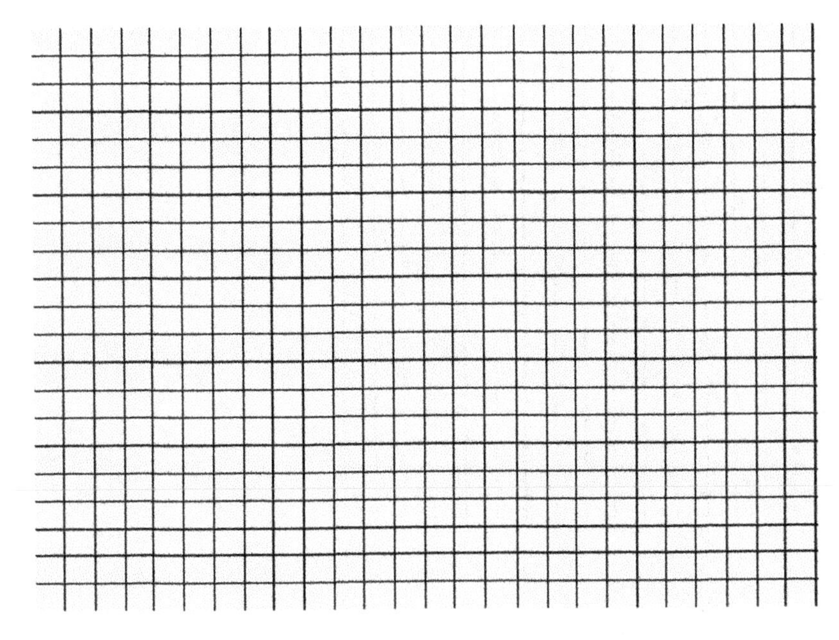

c. $f(x) = (x-1)(x-2)(x+3)(x+4)(x+4)$

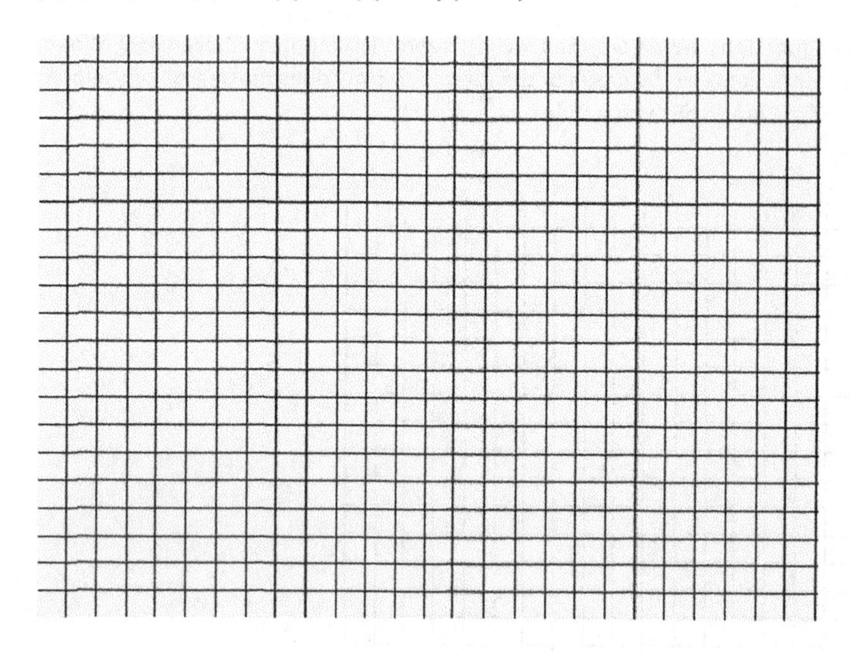

d. $f(x) = (x^2+1)(x-2)(x-3)$

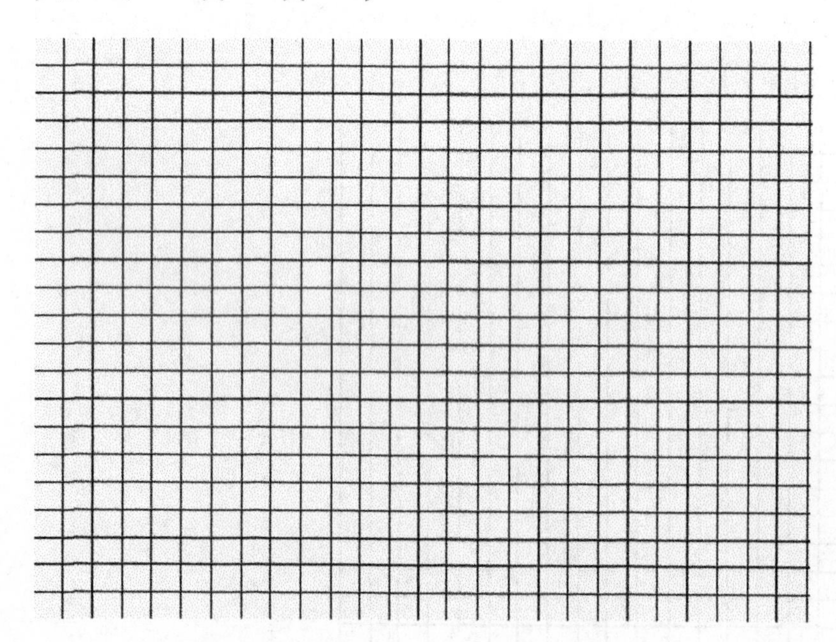

Example 2

Consider the function $f(x) = x^3 - 13x^2 + 44x - 32$.

a. Use the fact that $x - 4$ is a factor of f to factor this polynomial.

b. Find the x-intercepts for the graph of f.

c. At which x-values can the function change from being positive to negative or from negative to positive?

d. To sketch a graph of f, we need to consider whether the function is positive or negative on the four intervals $x < 1$, $1 < x < 4$, $4 < x < 8$, and $x > 8$. Why is that?

e. How can we tell if the function is positive or negative on an interval between x-intercepts?

f. For $x < 1$, is the graph above or below the x-axis? How can you tell?

g. For $1 < x < 4$, is the graph above or below the x-axis? How can you tell?

h. For $4 < x < 8$, is the graph above or below the x-axis? How can you tell?

i. For $x > 8$, is the graph above or below the x-axis? How can you tell?

j. Use the information generated in parts (f)–(i) to sketch a graph of f.

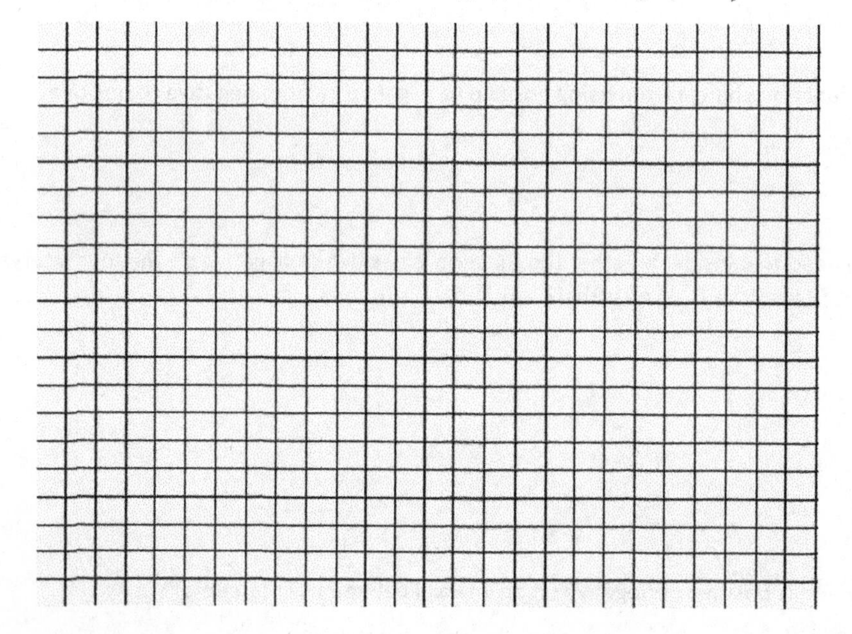

k. Graph $y = f(x)$ on the interval from $[0,9]$ using a graphing utility, and compare your sketch with the graph generated by the graphing utility.

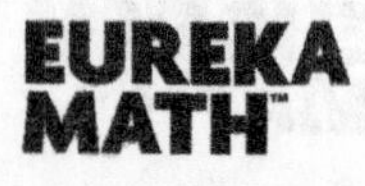

Discussion

For any particular polynomial, can we determine how many relative maxima or minima there are? Consider the following polynomial functions in factored form and their graphs.

$f(x) = (x + 1)(x - 3)$

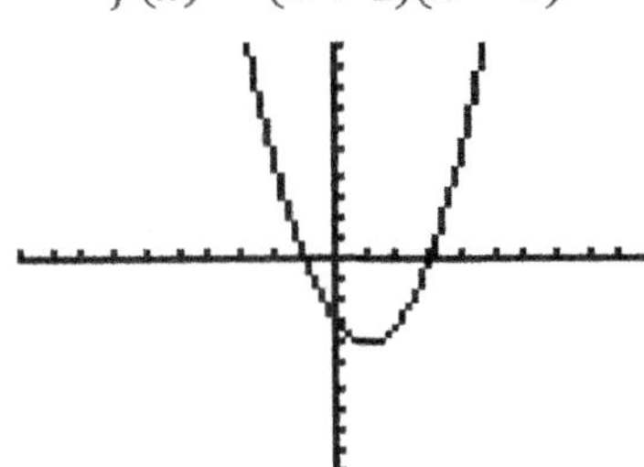

$g(x) = (x + 3)(x - 1)(x - 4)$

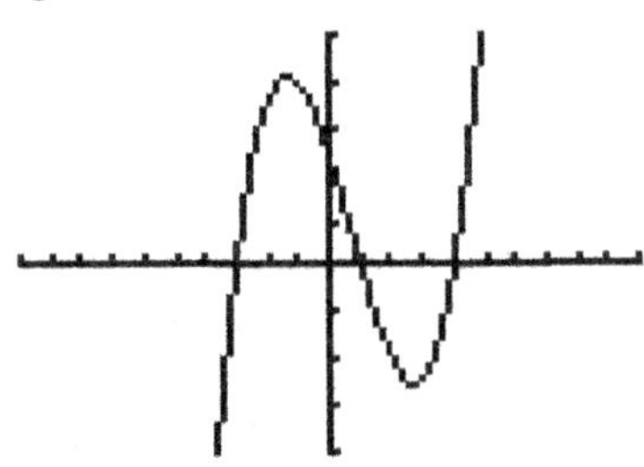

$h(x) = (x)(x + 4)(x - 2)(x - 5)$

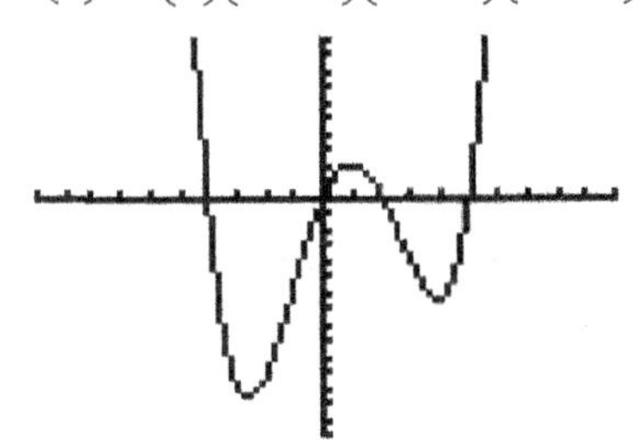

Degree of each polynomial:

Number of x-intercepts in each graph:

Number of relative maximum and minimum points shown in each graph:

What observations can we make from this information?

Is this true for every polynomial? Consider the examples below.

$r(x) = x^2 + 1$

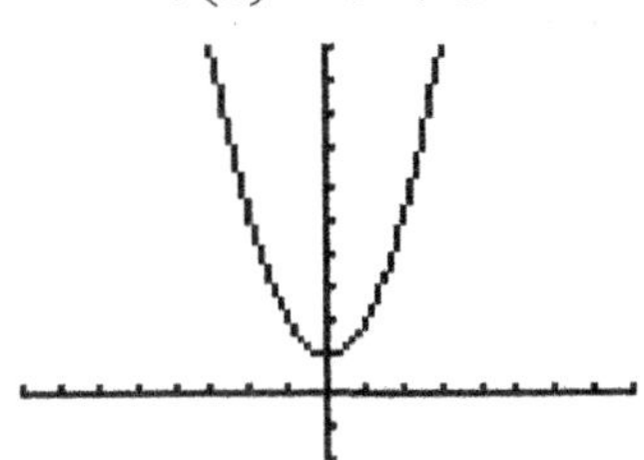

$s(x) = (x^2 + 2)(x - 1)$

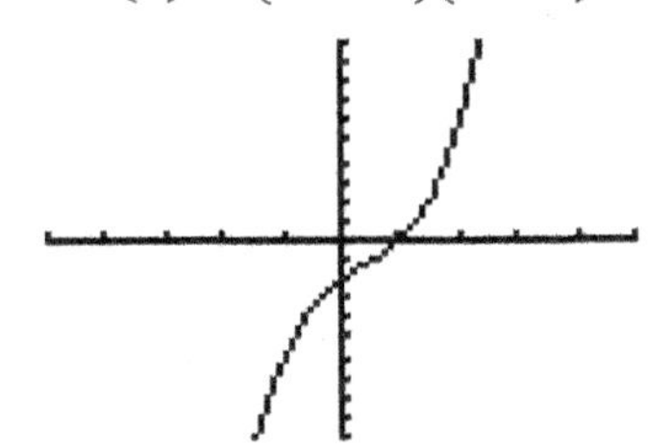

$t(x) = (x + 3)(x - 1)(x - 1)(x - 1)$

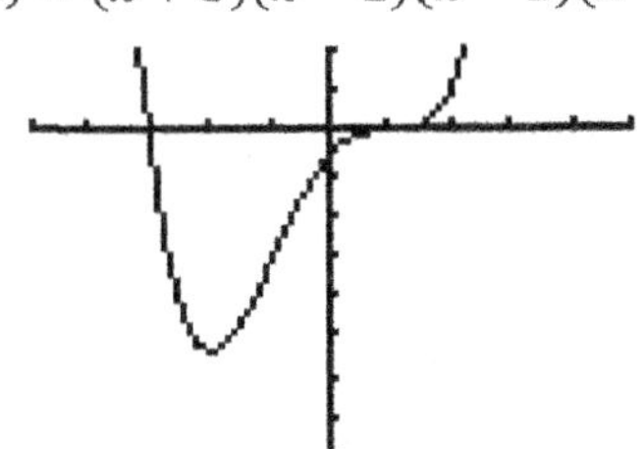

Degree of each polynomial:

Number of x-intercepts in each graph:

Number of relative maximum and minimum points shown in each graph:

What observations can we make from this information?

Relevant Vocabulary

INCREASING/DECREASING: Given a function f whose domain and range are subsets of the real numbers and I is an interval contained within the domain, the function is called *increasing on the interval* I if

$$f(x_1) < f(x_2) \text{ whenever } x_1 < x_2 \text{ in } I.$$

It is called *decreasing on the interval* I if

$$f(x_1) > f(x_2) \text{ whenever } x_1 < x_2 \text{ in } I.$$

RELATIVE MAXIMUM: Let f be a function whose domain and range are subsets of the real numbers. The function has a *relative maximum at* c if there exists an open interval I of the domain that contains c such that

$$f(x) \le f(c) \text{ for all } x \text{ in the interval } I.$$

If f has a relative maximum at c, then the value $f(c)$ is called the *relative maximum value*.

RELATIVE MINIMUM: Let f be a function whose domain and range are subsets of the real numbers. The function has a *relative minimum at* c if there exists an open interval I of the domain that contains c such that

$$f(x) \ge f(c) \text{ for all } x \text{ in the interval } I.$$

If f has a relative minimum at c, then the value $f(c)$ is called the *relative minimum value*.

GRAPH OF f: Given a function f whose domain D and the range are subsets of the real numbers, the graph of f is the set of ordered pairs in the Cartesian plane given by

$$\{(x, f(x)) \mid x \in D\}.$$

GRAPH OF $y = f(x)$: Given a function f whose domain D and the range are subsets of the real numbers, the *graph of* $y = f(x)$ is the set of ordered pairs (x, y) in the Cartesian plane given by

$$\{(x, y) \mid x \in D \text{ and } y = f(x)\}.$$

Lesson Summary

A polynomial of degree n may have up to n x-intercepts and up to $n-1$ relative maximum/minimum points.

The function f has a relative maximum at c if there is an open interval around c so that for all x in that interval, $f(x) \leq f(c)$. That is, looking near the point $(c, f(c))$ on the graph of f, there is no point higher than $(c, f(c))$ in that region. The value $f(c)$ is a relative maximum value.

The function f has a relative minimum at d if there is an open interval around d so that for all x in that interval, $f(x) \geq f(d)$. That is, looking near the point $(d, f(d))$ on the graph of f, there is no point lower than $(d, f(d))$ in that region. The value $f(d)$ is a relative minimum value.

The plural of maximum is maxima, and the plural of minimum is minima.

Problem Set

1. For each function below, identify the largest possible number of x-intercepts and the largest possible number of relative maxima and minima based on the degree of the polynomial. Then use a calculator or graphing utility to graph the function and find the actual number of x-intercepts and relative maxima and minima.
   a. $f(x) = 4x^3 - 2x + 1$
   b. $g(x) = x^7 - 4x^5 - x^3 + 4x$
   c. $h(x) = x^4 + 4x^3 + 2x^2 - 4x + 2$

Function	Largest number of x-intercepts	Largest number of relative max/min	Actual number of x-intercepts	Actual number of relative max/min
a. f				
b. g				
c. h				

2. Sketch a graph of the function $f(x) = \frac{1}{2}(x + 5)(x + 1)(x - 2)$ by finding the zeros and determining the sign of the values of the function between zeros.

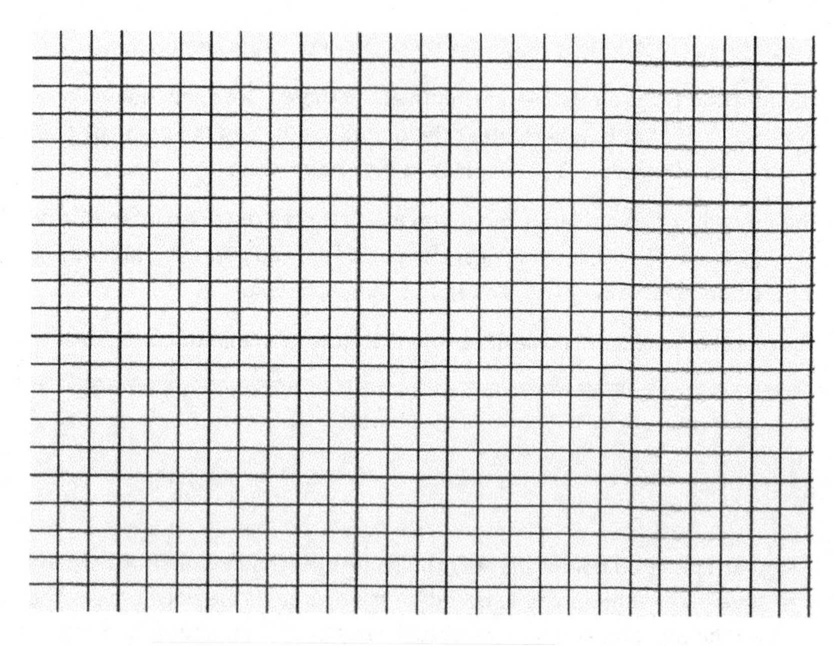

3. Sketch a graph of the function $f(x) = -(x + 2)(x - 4)(x - 6)$ by finding the zeros and determining the sign of the values of the function between zeros.

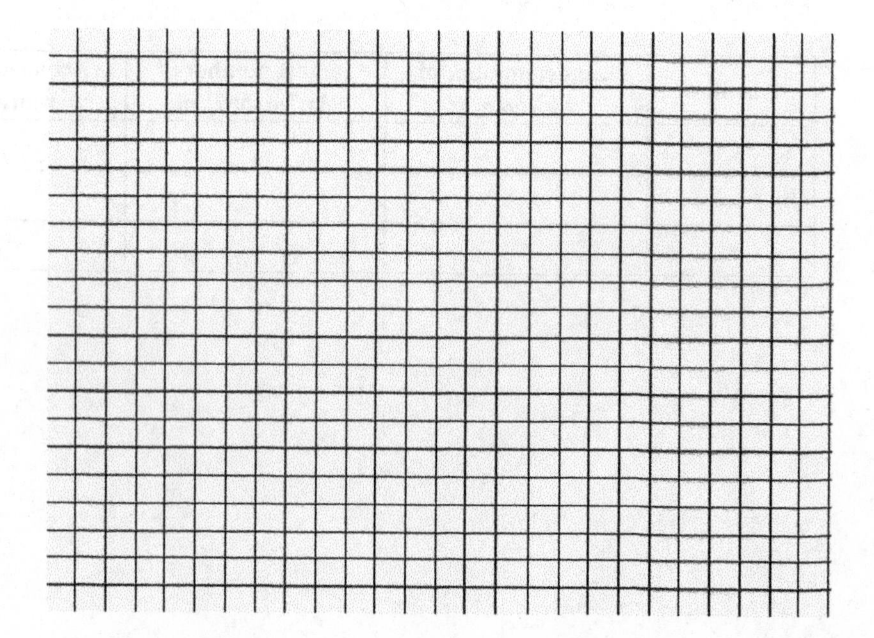

4. Sketch a graph of the function $f(x) = x^3 - 2x^2 - x + 2$ by finding the zeros and determining the sign of the values of the function between zeros.

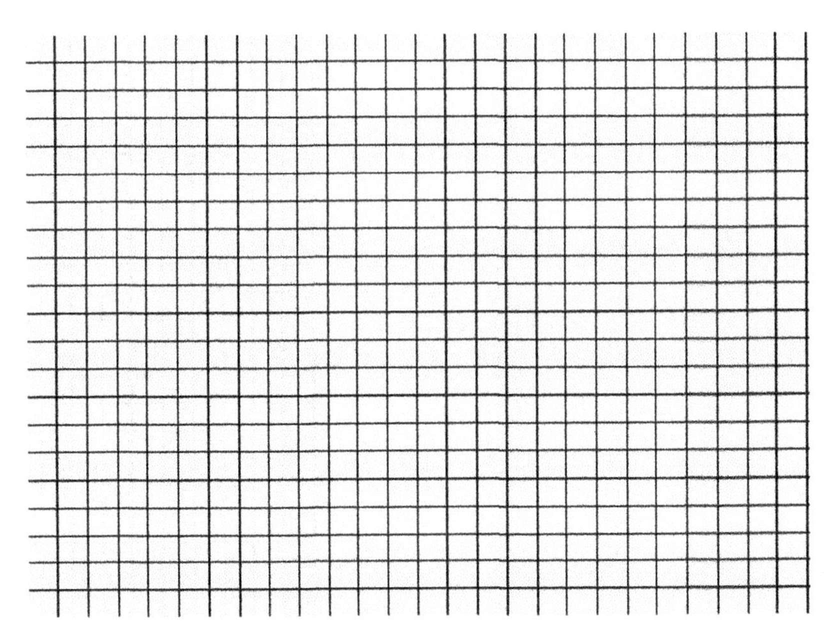

5. Sketch a graph of the function $f(x) = x^4 - 4x^3 + 2x^2 + 4x - 3$ by determining the sign of the values of the function between the zeros -1, 1, and 3.

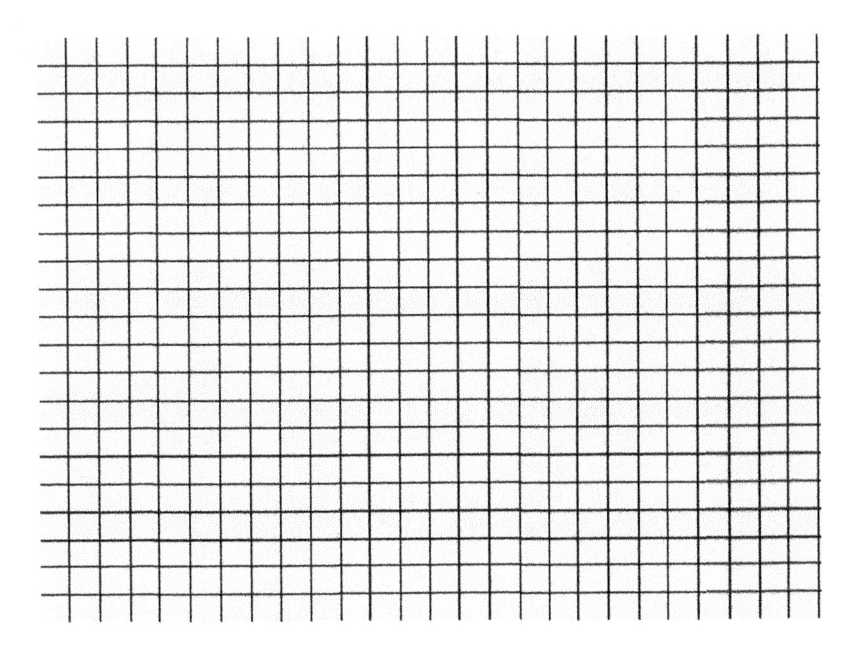

6. A function f has zeros at -1, 3, and 5. We know that $f(-2)$ and $f(2)$ are negative, while $f(4)$ and $f(6)$ are positive. Sketch a graph of f.

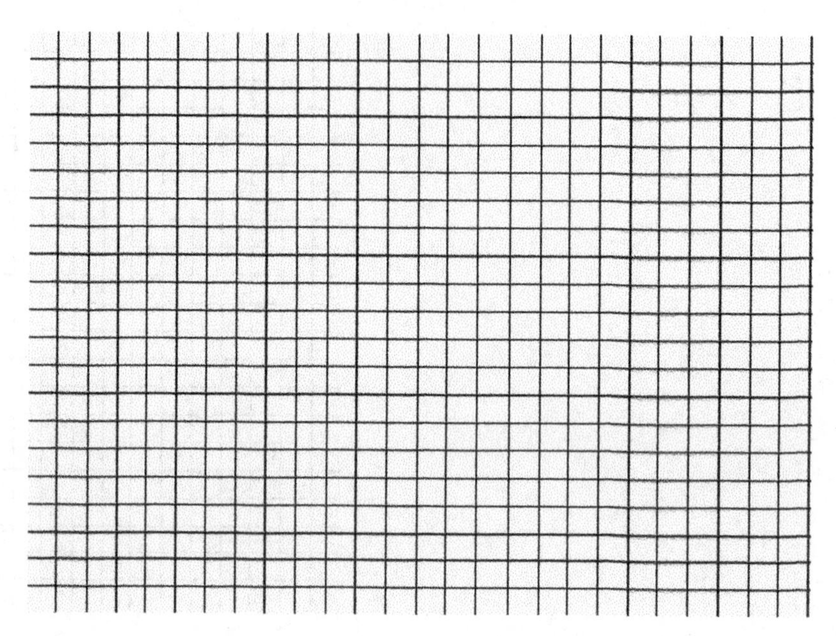

7. The function $h(t) = -16t^2 + 33t + 45$ represents the height of a ball tossed upward from the roof of a building 45 feet in the air after t seconds. Without graphing, determine when the ball will hit the ground.

Lesson 15: Structure in Graphs of Polynomial Functions

Classwork

Opening Exercise

Sketch the graph of $f(x) = x^2$. What will the graph of $g(x) = x^4$ look like? Sketch it on the same coordinate plane. What will the graph of $h(x) = x^6$ look like?

Example 1

Sketch the graph of $f(x) = x^3$. What will the graph of $g(x) = x^5$ look like? Sketch this on the same coordinate plane. What will the graph of $h(x) = x^7$ look like? Sketch this on the same coordinate plane.

Exercise 1

a. Consider the following function, $f(x) = 2x^4 + x^3 - x^2 + 5x + 3$, with a mixture of odd and even degree terms. Predict whether its end behavior will be like the functions in the Opening Exercise or more like the functions from Example 1. Graph the function f using a graphing utility to check your prediction.

b. Consider the following function, $f(x) = 2x^5 - x^4 - 2x^3 + 4x^2 + x + 3$, with a mixture of odd and even degree terms. Predict whether its end behavior will be like the functions in the Opening Exercise or more like the functions from Example 1. Graph the function f using a graphing utility to check your prediction.

c. Thinking back to our discussion of x-intercepts of graphs of polynomial functions from the previous lesson, sketch a graph of an even-degree polynomial function that has no x-intercepts.

d. Similarly, can you sketch a graph of an odd-degree polynomial function with no x-intercepts?

Exercise 2

The Center for Transportation Analysis (CTA) studies all aspects of transportation in the United States, from energy and environmental concerns to safety and security challenges. A 1997 study compiled the following data of the fuel economy in miles per gallon (mpg) of a car or light truck at various speeds measured in miles per hour (mph). The data are compiled in the table below.

Fuel Economy by Speed

Speed (mph)	Fuel Economy (mpg)
15	24.4
20	27.9
25	30.5
30	31.7
35	31.2
40	31.0
45	31.6
50	32.4
55	32.4
60	31.4
65	29.2
70	26.8
75	24.8

Source: Transportation Energy Data Book, Table 4.28. http://cta.ornl.gov/data/chapter4.shtml

a. Plot the data using a graphing utility. Which variable is the independent variable?

b. This data can be modeled by a polynomial function. Determine if the function that models the data would have an even or odd degree.

c. Is the leading coefficient of the polynomial that can be used to model this data positive or negative?

d. List two possible reasons the data might have the shape that it does.

Relevant Vocabulary

EVEN FUNCTION: Let f be a function whose domain and range is a subset of the real numbers. The function f is called *even* if the equation $f(x) = f(-x)$ is true for every number x in the domain.

Even-degree polynomial functions are sometimes even functions, like $f(x) = x^{10}$, and sometimes not, like $g(x) = x^2 - x$.

ODD FUNCTION: Let f be a function whose domain and range is a subset of the real numbers. The function f is called *odd* if the equation $f(-x) = -f(x)$ is true for every number x in the domain.

Odd-degree polynomial functions are sometimes odd functions, like $f(x) = x^{11}$, and sometimes not, like $h(x) = x^3 - x^2$.

Problem Set

1. Graph the functions from the Opening Exercise simultaneously using a graphing utility and zoom in at the origin.
   a. At $x = 0.5$, order the values of the functions from least to greatest.
   b. At $x = 2.5$, order the values of the functions from least to greatest.
   c. Identify the x-value(s) where the order reverses. Write a brief sentence on why you think this switch occurs.

2. The National Agricultural Statistics Service (NASS) is an agency within the USDA that collects and analyzes data covering virtually every aspect of agriculture in the United States. The following table contains information on the amount (in tons) of the following vegetables produced in the U.S. from 1988–1994 for processing into canned, frozen, and packaged foods: lima beans, snap beans, beets, cabbage, sweet corn, cucumbers, green peas, spinach, and tomatoes.

Vegetable Production by Year

Year	Vegetable Production (tons)
1988	11,393,320
1989	14,450,860
1990	15,444,970
1991	16,151,030
1992	14,236,320
1993	14,904,750
1994	18,313,150

Source: NASS Statistics of Vegetables and Melons, 1995, Table 191.
http://www.nass.usda.gov/Publications/Ag_Statistics/1995-1996/agr95_4.pdf

   a. Plot the data using a graphing utility.
   b. Determine if the data display the characteristics of an odd- or even-degree polynomial function.
   c. List two possible reasons the data might have such a shape.

3. The U.S. Energy Information Administration (EIA) is responsible for collecting and analyzing information about energy production and use in the United States and for informing policy makers and the public about issues of energy, the economy, and the environment. The following table contains data from the EIA about natural gas consumption from 1950–2010, measured in millions of cubic feet.

U.S. Natural Gas Consumption by Year

Year	U.S. natural gas total consumption (millions of cubic feet)
1950	5.77
1955	8.69
1960	11.97
1965	15.28
1970	21.14
1975	19.54
1980	19.88
1985	17.28
1990	19.17
1995	22.21
2000	23.33
2005	22.01
2010	24.09

Source: U.S. Energy Information Administration. http://www.eia.gov/dnav/ng/hist/n9140us2a.htm

a. Plot the data using a graphing utility.

b. Determine if the data display the characteristics of an odd- or even-degree polynomial function.

c. List two possible reasons the data might have such a shape.

4. We use the term *even function* when a function f satisfies the equation $f(-x) = f(x)$ for every number x in its domain. Consider the function $f(x) = -3x^2 + 7$. Note that the degree of the function is even, and each term is of an even degree (the constant term is degree 0.

a. Graph the function using a graphing utility.

b. Does this graph display any symmetry?

c. Evaluate $f(-x)$.

d. Is f an even function? Explain how you know.

5. We use the term *odd function* when a function f satisfies the equation $f(-x) = -f(x)$ for every number x in its domain. Consider the function $f(x) = 3x^3 - 4x$. The degree of the function is odd, and each term is of an odd degree.

a. Graph the function using a graphing utility.

b. Does this graph display any symmetry?

c. Evaluate $f(-x)$.

d. Is f an odd function? Explain how you know.

6. We have talked about x-intercepts of the graph of a function in both this lesson and the previous one. The x-intercepts correspond to the zeros of the function. Consider the following examples of polynomial functions and their graphs to determine an easy way to find the y-intercept of the graph of a polynomial function.

$f(x) = 2x^2 - 4x - 3$

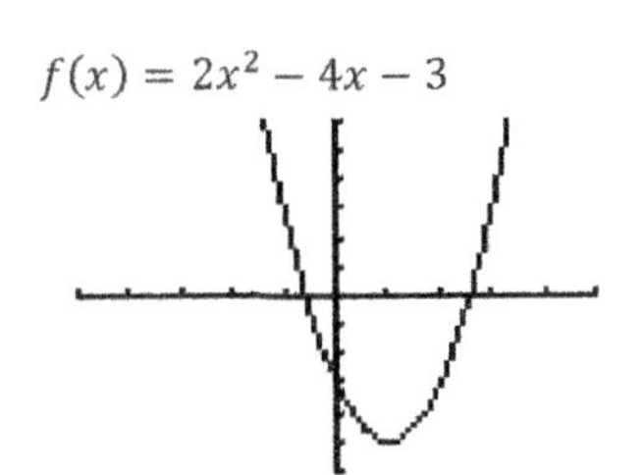

$f(x) = x^3 + 3x^2 - x + 5$

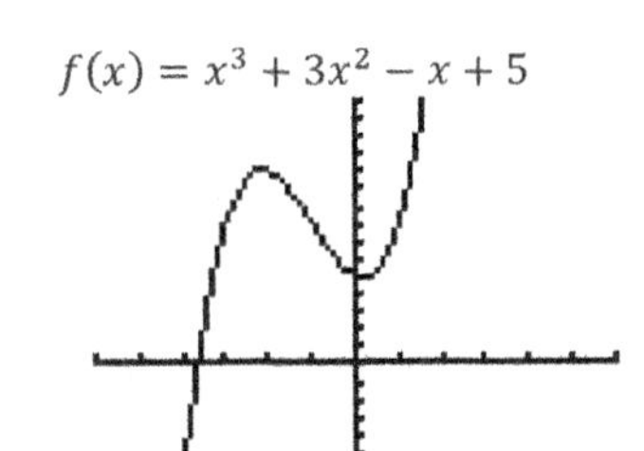

$f(x) = x^4 - 2x^3 - x^2 + 3x - 6$

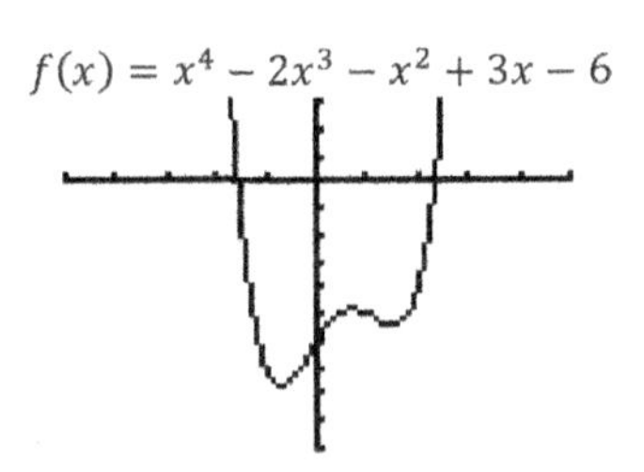

This page intentionally left blank

Lesson 16: Modeling with Polynomials—An Introduction

Classwork

Mathematical Modeling Exercise

You will be assigned to a group, which will create a box from a piece of construction paper. Each group will record its box's measurements and use said measurement values to calculate and record the volume of its box. Each group will contribute to the following class table on the board.

Group	Length	Width	Height	Volume
1				
2				
3				
4				

Using the given construction paper, cut out congruent squares from each corner, and fold the sides in order to create an open-topped box as shown on the figure below.

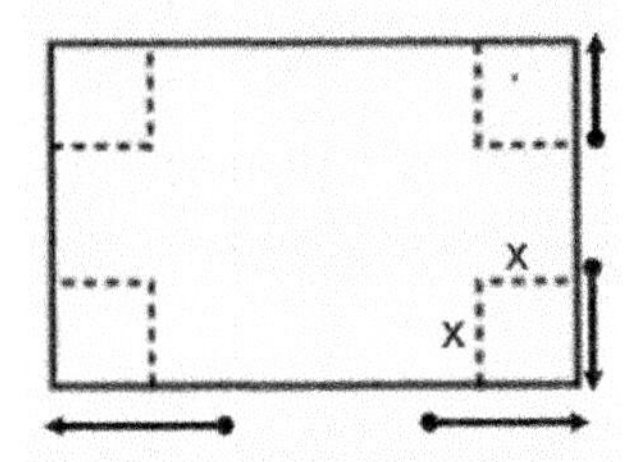

1. Measure the length, width, and height of the box to the nearest tenth of a centimeter.

2. Calculate the volume.

3. Have a group member record the values on the table on the board.

4. Create a scatterplot of volume versus height using technology.

5. What type of polynomial function could we use to model the data?

6. Use the regression feature to find a function to model the data. Does a quadratic or a cubic regression provide a better fit to the data?

7. Find the maximum volume of the box.

8. What size square should be cut from each corner in order to maximize the volume?

9. What are the dimensions of the box of maximum volume?

Problem Set

1. For a fundraiser, members of the math club decide to make and sell "Pythagoras may have been Fermat's first problem but not his last" t-shirts. They are trying to decide how many t-shirts to make and sell at a fixed price. They surveyed the level of interest of students around school and made a scatterplot of the number of t-shirts sold (x) versus profit shown below.

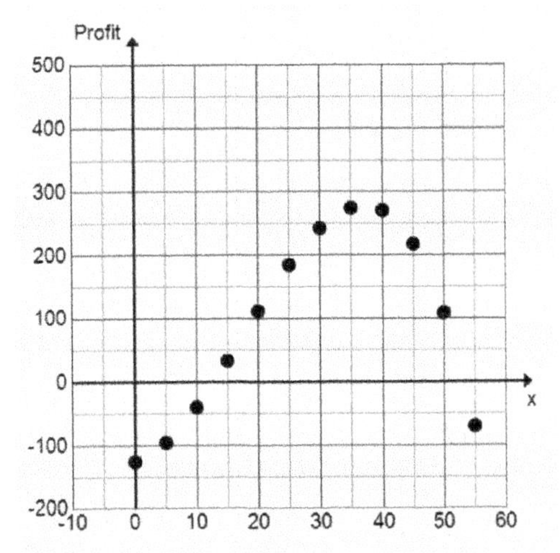

   a. Identify the y-intercept. Interpret its meaning within the context of this problem.
   b. If we model this data with a function, what point on the graph of that function represents the number of t-shirts they need to sell in order to break even? Why?
   c. What is the smallest number of t-shirts they can sell and still make a profit?
   d. How many t-shirts should they sell in order to maximize the profit?
   e. What is the maximum profit?
   f. What factors would affect the profit?
   g. What would cause the profit to start decreasing?

2. The following graph shows the temperature in Aspen, Colorado during a 48-hour period beginning at midnight on Thursday, January 21, 2014. (Source: National Weather Service)

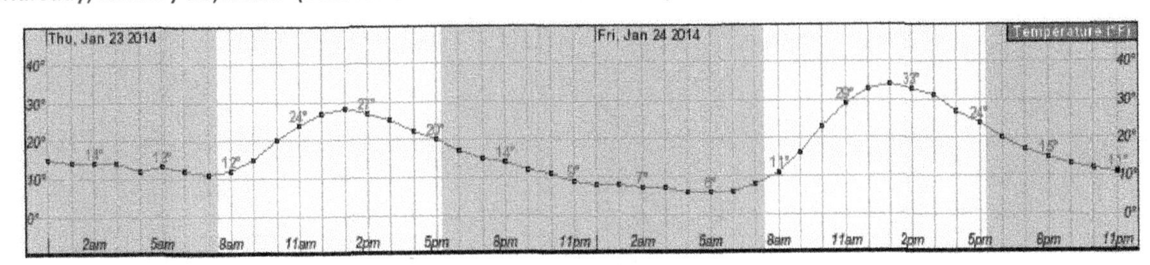

   a. We can model the data shown with a polynomial function. What degree polynomial would be a reasonable choice?
   b. Let T be the function that represents the temperature, in degrees Fahrenheit, as a function of time t, in hours. If we let $t = 0$ correspond to midnight on Thursday, interpret the meaning of $T(5)$. What is $T(5)$?
   c. What are the relative maximum values? Interpret their meanings.

This page intentionally left blank

Lesson 17: Modeling with Polynomials—An Introduction

Classwork

Opening Exercise

In Lesson 16, we created an open-topped box by cutting congruent squares from each corner of a piece of construction paper.

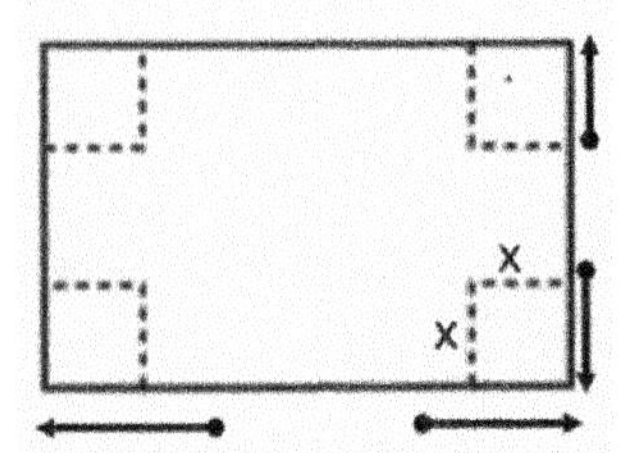

a. Express the dimensions of the box in terms of x.

b. Write a formula for the volume of the box as a function of x. Give the answer in standard form.

Mathematical Modeling Exercises 1–13

The owners of Dizzy Lizzy's, an amusement park, are studying the wait time at their most popular roller coaster. The table below shows the number of people standing in line for the roller coaster t hours after Dizzy Lizzy's opens.

t (hours)	0	1	2	4	7	8	10	12
P (people in line)	0	75	225	345	355	310	180	45

Jaylon made a scatterplot and decided that a cubic function should be used to model the data. His scatterplot and curve are shown below.

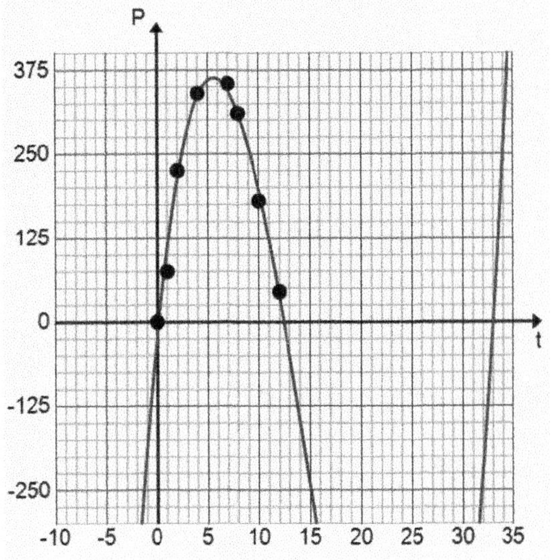

1. Do you agree that a cubic polynomial function is a good model for this data? Explain.

2. What information would Dizzy Lizzy's be interested in learning about from this graph? How could they determine the answer?

3. Estimate the time at which the line is the longest. Explain how you know.

4. Estimate the number of people in line at that time. Explain how you know.

5. Estimate the t-intercepts of the function used to model this data.

6. Use the t-intercepts to write a formula for the function of the number of people in line, f, after t hours.

7. Use the relative maximum to find the leading coefficient of f. Explain your reasoning.

8. What would be a reasonable domain for your function f? Why?

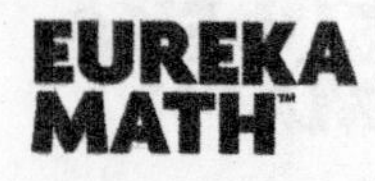

9. Use your function f to calculate the number of people in line 10 hours after the park opens.

10. Comparing the value calculated above to the actual value in the table, is your function f an accurate model for the data? Explain.

11. Use the regression feature of a graphing calculator to find a cubic function g to model the data.

12. Graph the function f you found and the function g produced by the graphing calculator. Use the graphing calculator to complete the table. Round your answers to the nearest integer.

t (hours)	0	1	2	4	7	8	10	12
P (people in line)	0	75	225	345	355	310	180	45
$f(t)$ (your equation)								
$g(t)$ (regression eqn.)								

13. Based on the results from the table, which model was more accurate at $t = 2$ hours? $t = 10$ hours?

Problem Set

1. Recall the math club fundraiser from the Problem Set of the previous lesson. The club members would like to find a function to model their data, so Kylie draws a curve through the data points as shown.

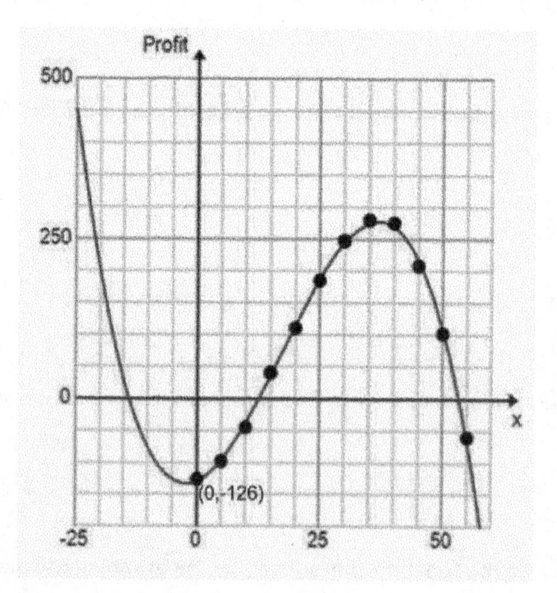

   a. What type of function does it appear that she has drawn?
   b. The function that models the profit in terms of the number of t-shirts made has the form $P(x) = c(x^3 - 53x^2 - 236x + 9828)$. Use the vertical intercept labeled on the graph to find the value of the leading coefficient c.
   c. From the graph, estimate the profit if the math club sells 30 t-shirts.
   d. Use your function to estimate the profit if the math club sells 30 t-shirts.
   e. Which estimate do you think is more reliable? Why?

2. A box is to be constructed so that it has a square base and no top.
   a. Draw and label the sides of the box. Label the sides of the base as x and the height of the box as h.
   b. The surface area is 108 cm^2. Write a formula for the surface area S, and then solve for h.
   c. Write a formula for the function of the volume of the box in terms of x.
   d. Use a graphing utility to find the maximum volume of the box.
   e. What dimensions should the box be in order to maximize its volume?

Lesson 18: Overcoming a Second Obstacle in Factoring—What If There Is a Remainder?

Classwork

Opening Exercise

Write the rational number $\frac{13}{4}$ as a mixed number.

Example 1

a. Find the quotient by factoring the numerator.

$$\frac{x^2 + 3x + 2}{x + 2}$$

b. Find the quotient.

$$\frac{x^2 + 3x + 5}{x + 2}$$

Example 2

a. Find the quotient by factoring the numerator.

$$\frac{x^3 - 8}{x - 2}$$

b. Find the quotient.

$$\frac{x^3 - 4}{x - 2}$$

Exercises 1–10

Find each quotient by inspection.

1. $\dfrac{x + 4}{x + 1}$

2. $\dfrac{2x - 7}{x - 3}$

3. $\dfrac{x^2 - 21}{x + 4}$

Find each quotient by using the reverse tabular method.

4. $\dfrac{x^2 + 4x + 10}{x - 8}$

5. $\dfrac{x^3 - x^2 + 3x - 1}{x + 3}$

6. $\dfrac{x^2 - 2x - 19}{x - 1}$

Find each quotient by using long division.

7. $\dfrac{x^2 - x - 25}{x + 6}$

8. $\dfrac{x^4 - 8x^2 + 12}{x + 2}$

9. $\dfrac{4x^3 + 5x - 8}{2x - 5}$

Rewrite the numerator in the form $(x-h)^2+k$ by completing the square. Then find the quotient.

10. $\dfrac{x^2+4x-9}{x+2}$

Mental Math Exercises

$\dfrac{x^2-9}{x+3}$	$\dfrac{x^2-4x+3}{x-1}$	$\dfrac{x^2-16}{x+4}$	$\dfrac{x^2-3x-4}{x+1}$
$\dfrac{x^3-3x^2}{x-3}$	$\dfrac{x^4-x^2}{x^2-1}$	$\dfrac{x^2+x-6}{x+3}$	$\dfrac{x^2-4}{x+2}$
$\dfrac{x^2-8x+12}{x-2}$	$\dfrac{x^2-36}{x+6}$	$\dfrac{x^2+6x+8}{x+4}$	$\dfrac{x^2-4}{x-2}$
$\dfrac{x^2-x-20}{x+4}$	$\dfrac{x^2-25}{x+5}$	$\dfrac{x^2-2x+1}{x-1}$	$\dfrac{x^2-3x+2}{x-2}$
$\dfrac{x^2+4x-5}{x-1}$	$\dfrac{x^2-25}{x-5}$	$\dfrac{x^2-10x}{x}$	$\dfrac{x^2-12x+20}{x-2}$
$\dfrac{x^2+5x+4}{x+4}$	$\dfrac{x^2-1}{x-1}$	$\dfrac{x^2+16x+64}{x+8}$	$\dfrac{x^2+9x+8}{x+1}$

Problem Set

1. For each pair of problems, find the first quotient by factoring the numerator. Then, find the second quotient by using the first quotient.

   a. $\dfrac{3x-6}{x-2}$ $\dfrac{3x-9}{x-2}$

   b. $\dfrac{x^2-5x-14}{x-7}$ $\dfrac{x^2-5x+2}{x-7}$

   c. $\dfrac{x^3+1}{x+1}$ $\dfrac{x^3}{x+1}$

   d. $\dfrac{x^2-13x+36}{x-4}$ $\dfrac{x^2-13x+30}{x-4}$

Find each quotient by using the reverse tabular method.

2. $\dfrac{x^3-9x^2+5x+2}{x-1}$

3. $\dfrac{x^2+x+10}{x+12}$

4. $\dfrac{2x+6}{x-8}$

5. $\dfrac{x^2+8}{x+3}$

Find each quotient by using long division.

6. $\dfrac{x^4-9x^2+10x}{x+2}$

7. $\dfrac{x^5-35}{x-2}$

8. $\dfrac{x^2}{x-6}$

9. $\dfrac{x^3+2x^2+8x+1}{x+5}$

10. $\dfrac{x^3+2x+11}{x-1}$

11. $\dfrac{x^4+3x^3-2x^2+6x-15}{x}$

12. Rewrite the numerator in the form $(x-h)^2+k$ by completing the square. Then, find the quotient.

$$\frac{x^2-6x-10}{x-3}$$

Lesson 19: The Remainder Theorem

Classwork

Exercises 1–3

1. Consider the polynomial function $f(x) = 3x^2 + 8x - 4$.

   a. Divide f by $x - 2$.

   b. Find $f(2)$.

2. Consider the polynomial function $g(x) = x^3 - 3x^2 + 6x + 8$.

   a. Divide g by $x + 1$.

   b. Find $g(-1)$.

3. Consider the polynomial function $h(x) = x^3 + 2x - 3$.

   a. Divide h by $x - 3$.

   b. Find $h(3)$.

Exercises 4–6

4. Consider the polynomial $P(x) = x^3 + kx^2 + x + 6$.

   a. Find the value of k so that $x + 1$ is a factor of P.

   b. Find the other two factors of P for the value of k found in part (a).

5. Consider the polynomial $P(x) = x^4 + 3x^3 - 28x^2 - 36x + 144$.

   a. Is 1 a zero of the polynomial P?

   b. Is $x + 3$ one of the factors of P?

   c. The graph of P is shown to the right. What are the zeros of P?

   d. Write the equation of P in factored form.

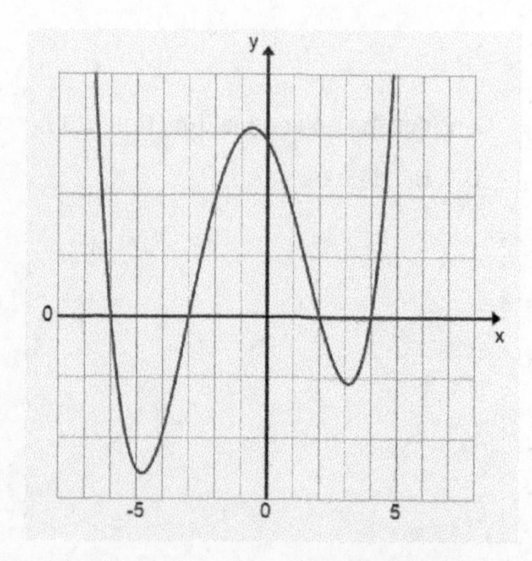

6. Consider the graph of a degree 5 polynomial shown to the right, with x-intercepts -4, -2, 1, 3, and 5.

   a. Write a formula for a possible polynomial function that the graph represents using c as the constant factor.

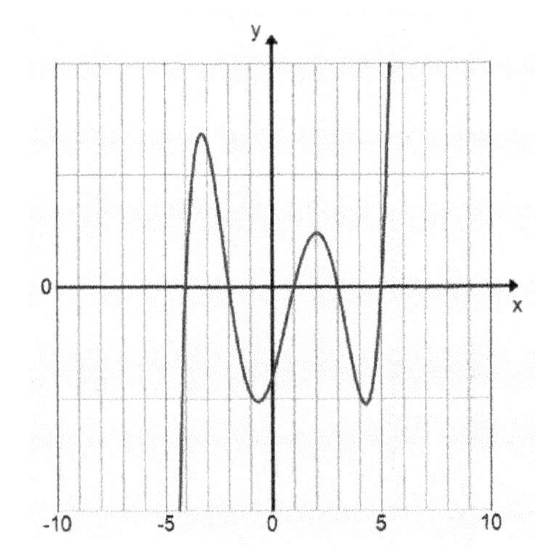

   b. Suppose the y-intercept is -4. Find the value of c so that the graph of P has y-intercept -4.

Lesson Summary

REMAINDER THEOREM: Let P be a polynomial function in x, and let a be any real number. Then there exists a unique polynomial function q such that the equation

$$P(x) = q(x)(x - a) + P(a)$$

is true for all x. That is, when a polynomial is divided by $(x - a)$, the remainder is the value of the polynomial evaluated at a.

FACTOR THEOREM: Let P be a polynomial function in x, and let a be any real number. If a is a zero of P, then $(x - a)$ is a factor of P.

Example: If $P(x) = x^2 - 3$ and $a = 4$, then $P(x) = (x + 4)(x - 4) + 13$ where $q(x) = x + 4$ and $P(4) = 13$.

Example: If $P(x) = x^3 - 5x^2 + 3x + 9$, then $P(3) = 27 - 45 + 9 + 9 = 0$, so $(x - 3)$ is a factor of P.

Problem Set

1. Use the remainder theorem to find the remainder for each of the following divisions.
   a. $\frac{(x^2+3x+1)}{(x+2)}$
   b. $\frac{x^3-6x^2-7x+9}{(x-3)}$
   c. $\frac{32x^4+24x^3-12x^2+2x+1}{(x+1)}$
   d. $\frac{32x^4+24x^3-12x^2+2x+1}{(2x-1)}$,

   Hint for part (d): Can you rewrite the division expression so that the divisor is in the form $(x - c)$ for some constant c?

2. Consider the polynomial $P(x) = x^3 + 6x^2 - 8x - 1$. Find $P(4)$ in two ways.

3. Consider the polynomial function $P(x) = 2x^4 + 3x^2 + 12$.
   a. Divide P by $x + 2$, and rewrite P in the form (divisor)(quotient) + remainder.
   b. Find $P(-2)$.

4. Consider the polynomial function $P(x) = x^3 + 42$.
   a. Divide P by $x - 4$, and rewrite P in the form (divisor)(quotient)+remainder.
   b. Find $P(4)$.

5. Explain why for a polynomial function P, $P(a)$ is equal to the remainder of the quotient of P and $x - a$.

6. Is $x - 5$ a factor of the function $f(x) = x^3 + x^2 - 27x - 15$? Show work supporting your answer.

7. Is $x + 1$ a factor of the function $f(x) = 2x^5 - 4x^4 + 9x^3 - x + 13$? Show work supporting your answer.

8. A polynomial function p has zeros of $2, 2, -3, -3, -3$, and 4. Find a possible formula for P, and state its degree. Why is the degree of the polynomial not 3?

9. Consider the polynomial function $P(x) = x^3 - 8x^2 - 29x + 180$.
   a. Verify that $P(9) = 0$. Since $P(9) = 0$, what must one of the factors of P be?
   b. Find the remaining two factors of P.
   c. State the zeros of P.
   d. Sketch the graph of P.

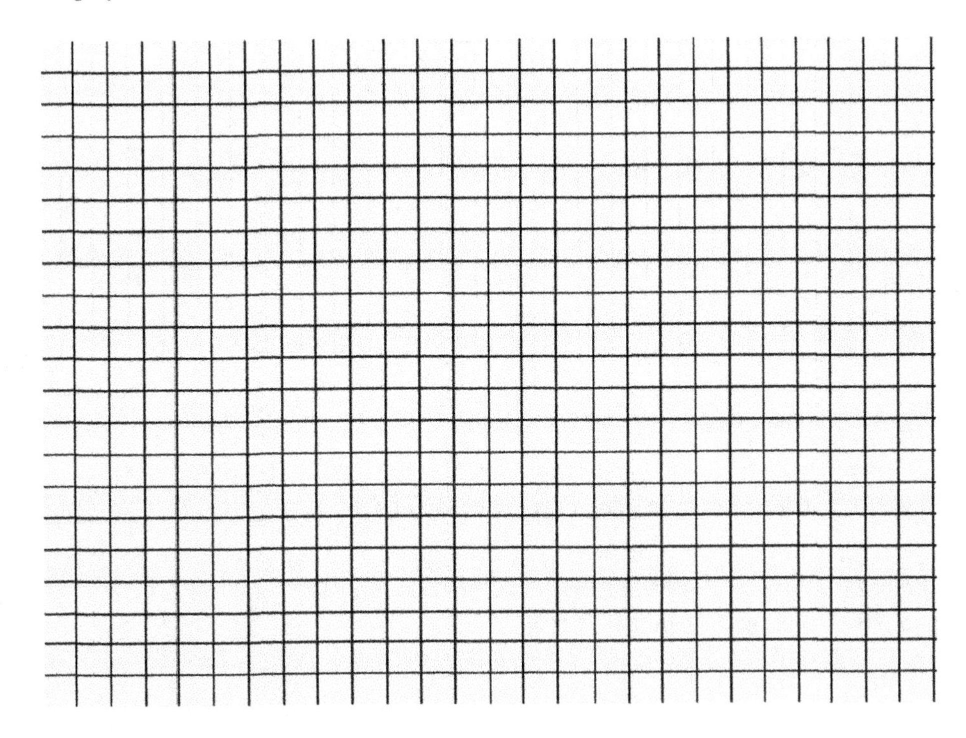

10. Consider the polynomial function $P(x) = 2x^3 + 3x^2 - 2x - 3$.
    a. Verify that $P(-1) = 0$. Since $P(-1) = 0$, what must one of the factors of P be?
    b. Find the remaining two factors of P.
    c. State the zeros of P.
    d. Sketch the graph of P.

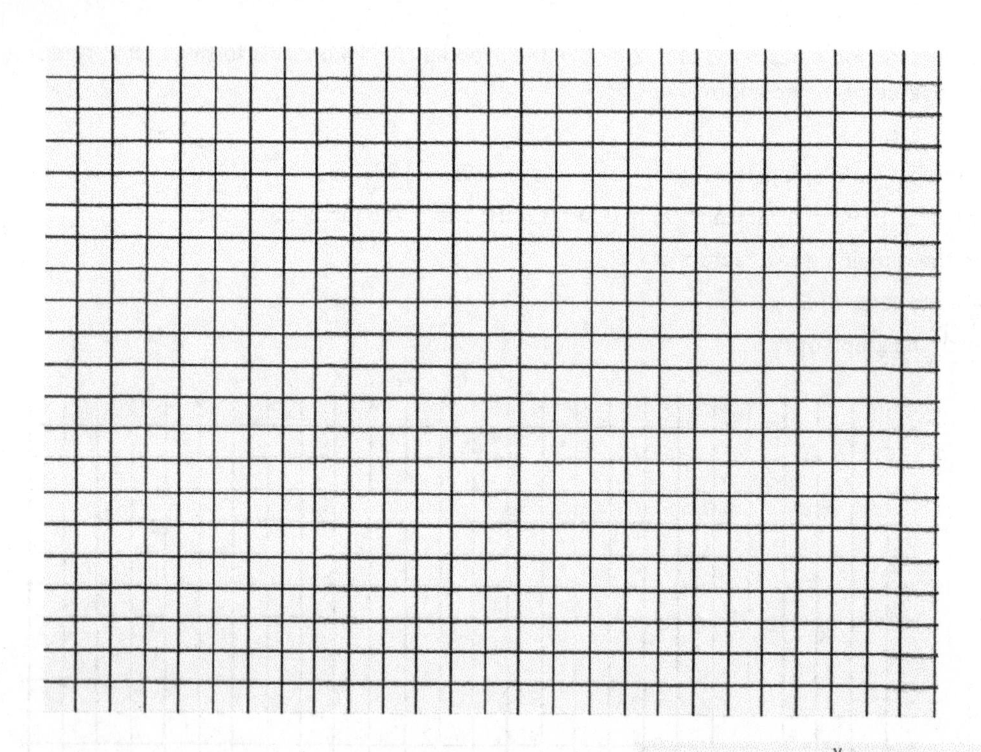

11. The graph to the right is of a third-degree polynomial function f.
    a. State the zeros of f.
    b. Write a formula for f in factored form using c for the constant factor.
    c. Use the fact that $f(-4) = -54$ to find the constant factor c.
    d. Verify your equation by using the fact that $f(1) = 11$.

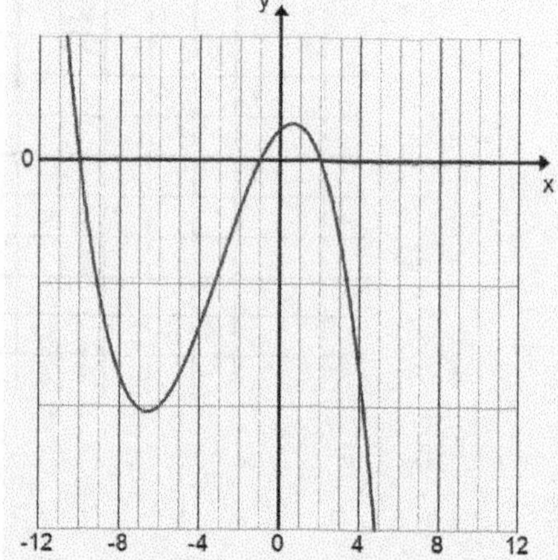

12. Find the value of k so that $\frac{x^3-kx^2+2}{x-1}$ has remainder 8.

13. Find the value k so that $\frac{kx^3+x-k}{x+2}$ has remainder 16.

14. Show that $x^{51} - 21x + 20$ is divisible by $x - 1$.

15. Show that $x + 1$ is a factor of $19x^{42} + 18x - 1$.

Write a polynomial function that meets the stated conditions.

16. The zeros are -2 and 1.

17. The zeros are -1, 2, and 7.

18. The zeros are $-\frac{1}{2}$ and $\frac{3}{4}$.

19. The zeros are $-\frac{2}{3}$ and 5, and the constant term of the polynomial is -10.

20. The zeros are 2 and $-\frac{3}{2}$, the polynomial has degree 3, and there are no other zeros.

This page intentionally left blank

Lesson 20: Modeling Riverbeds with Polynomials

Classwork

Mathematical Modeling Exercise

The Environmental Protection Agency (EPA) is studying the flow of a river in order to establish flood zones. The EPA hired a surveying company to determine the flow rate of the river, measured as volume of water per minute. The firm set up a coordinate system and found the depths of the river at five locations as shown on the graph below. After studying the data, the firm decided to model the riverbed with a polynomial function and divide the cross-sectional area into six regions that are either trapezoidal or triangular so that the overall area can be easily estimated. The firm needs to approximate the depth of the river at two more data points in order to do this.

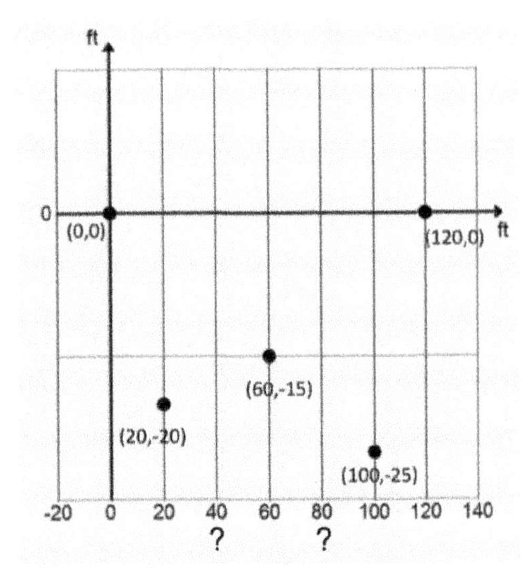

Draw the four trapezoids and two triangles that will be used to estimate the cross-sectional area of the riverbed.

Example 1

Find a polynomial P such that $P(0) = 28$, $P(2) = 0$, and $P(8) = 12$.

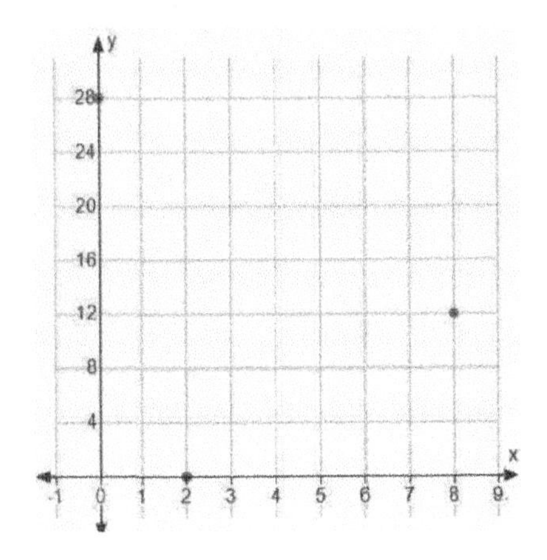

Example 2

Find a degree 3 polynomial P such that $P(-1) = -3$, $P(0) = -2$, $P(1) = -1$, and $P(2) = 6$.

Function Value	Substitute the data point into the current form of the equation for P.	Apply the remainder theorem to a, b, and c	Rewrite the equation for P in terms of a, b, or c.
$P(-1) = -3$			
$P(0) = -2$			
$P(1) = -1$			
$P(2) = 6$			

Lesson Summary

A linear polynomial is determined by 2 points on its graph.

A degree 2 polynomial is determined by 3 points on its graph.

A degree 3 polynomial is determined by 4 points on its graph.

A degree 4 polynomial is determined by 5 points on its graph.

The remainder theorem can be used to find a polynomial P whose graph will pass through a given set of points.

Problem Set

1. Suppose a polynomial function P is such that $P(2) = 5$ and $P(3) = 12$.
   a. What is the largest-degree polynomial that can be uniquely determined given the information?
   b. Is this the only polynomial that satisfies $P(2) = 5$ and $P(3) = 12$?
   c. Use the remainder theorem to find the polynomial P of least degree that satisfies the two points given.
   d. Verify that your equation is correct by demonstrating that it satisfies the given points.

2. Write a quadratic function P such that $P(0) = -10$, $P(5) = 0$, and $P(7) = 18$ using the specified method.
   a. Setting up a system of equations
   b. Using the remainder theorem

3. Find a degree-three polynomial function P such that $P(-1) = 0$, $P(0) = 2$, $P(2) = 12$, and $P(3) = 32$. Use the table below to organize your work. Write your answer in standard form, and verify by showing that each point satisfies the equation.

Function Value	Substitute the data point into the current form of the equation for P.	Apply the remainder theorem to a, b, and c.	Rewrite the equation for P in terms of a, b, or c.
$P(-1) = 0$			
$P(0) = 2$			
$P(2) = 12$			
$P(3) = 32$			

4. The method used in Problem 3 is based on the Lagrange interpolation method. Research Joseph-Louis Lagrange, and write a paragraph about his mathematical work.

This page intentionally left blank

Lesson 21: Modeling Riverbeds with Polynomials

Classwork

Mathematical Modeling Exercise

The Environmental Protection Agency (EPA) is studying the flow of a river in order to establish flood zones. The EPA hired a surveying company to determine the flow rate of the river, measured as volume of water per minute. The firm set up a coordinate system and found the depths of the river at five locations as shown on the graph below. After studying the data, the firm decided to model the riverbed with a polynomial function and divide the area into six regions that are either trapezoidal or triangular so that the overall area can be easily estimated. The firm needs to approximate the depth of the river at two more data points in order to do this.

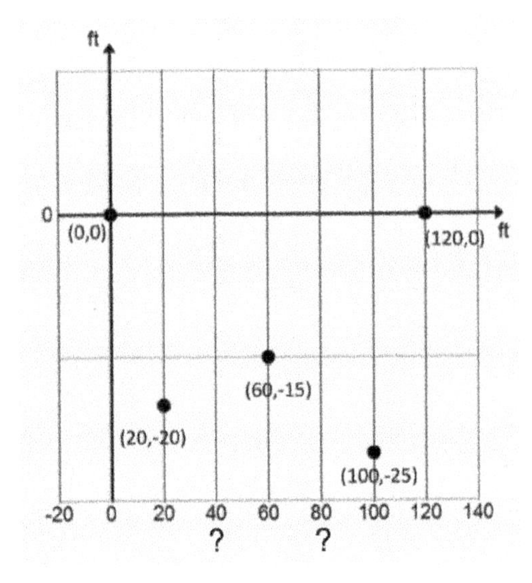

1. Find a polynomial P that fits the five given data points.

2. Use the polynomial to estimate the depth of the river at $x = 40$ and $x = 80$.

3. Estimate the area of the cross section.

Suppose that the river flow speed was measured to be an average speed of $176\frac{\text{ft.}}{\text{min}}$ the cross section.

4. What is the volumetric flow of the water (the volume of water per minute)?

5. Convert the flow to gallons per minute. [Note: 1 cubic foot ≈ 7.48052 gallons.]

Problem Set

1. As the leader of the surveying team, write a short report to the EPA on your findings from the in-class exercises. Be sure to include data and calculations.

2. Suppose that depths of the riverbed were measured for a different cross-section of the river.

   a. Use Wolfram Alpha to find the interpolating polynomial Q with values:

   $Q(0) = 0$, $Q(16.5) = -27.4$, $Q(44.4) = -19.6$, $Q(77.6) = -25.1$,
   $Q(123.3) = -15.0$, $Q(131.1) = -15.1$, $Q(150) = 0$.

   b. Sketch the cross-section of the river, and estimate its area.

   c. Suppose that the speed of the water was measured at $124\frac{\text{ft.}}{\text{min}}$. What is the approximate volumetric flow in this section of the river, measured in gallons per minute?

This page intentionally left blank

Lesson 22: Equivalent Rational Expressions

Classwork

Opening Exercise

On your own or with a partner, write two fractions that are equivalent to $\frac{1}{3}$, and use the slips of paper to create visual models to justify your response.

Example

Consider the following rational expression: $\frac{2(a-1)-2}{6(a-1)-3a}$. Turn to your neighbor, and discuss the following: For what values of a is the expression undefined?

Exercise

Reduce the following rational expressions to lowest terms, and identify the values of the variable(s) that must be excluded to prevent division by zero.

a. $\frac{2(x+1)+2}{(2x+3)(x+1)-1}$

b. $\frac{x^2-x-6}{5x^2+10x}$

c. $\frac{3-x}{x^2-9}$

d. $\frac{3x-3y}{y^2-2xy+x^2}$

EUREKA MATH™

Lesson Summary

- If a, b, and n are integers with $n \neq 0$ and $b \neq 0$, then
$$\frac{na}{nb} = \frac{a}{b}.$$
- The rule for rational expressions is the same as the rule for integers but requires the domain of the rational expression to be restricted (i.e., no value of the variable that can make the denominator of the original rational expression zero is allowed).

Problem Set

1. Find an equivalent rational expression in lowest terms, and identify the value(s) of the variable that must be excluded to prevent division by zero.

   a. $\frac{16n}{20n}$
   b. $\frac{x^3y}{y^4x}$
   c. $\frac{2n-8n^2}{4n}$
   d. $\frac{db+dc}{db}$
   e. $\frac{x^2-9b^2}{x^2-2xb-3b^2}$
   f. $\frac{3n^2-5n-2}{2n-4}$
   g. $\frac{3x-2y}{9x^2-4y^2}$
   h. $\frac{4a^2-12ab}{a^2-6ab+9b^2}$
   i. $\frac{y-x}{x-y}$
   j. $\frac{a^2-b^2}{b+a}$
   k. $\frac{4x-2y}{3y-6x}$
   l. $\frac{9-x^2}{(x-3)^3}$
   m. $\frac{x^2-5x+6}{8-2x-x^2}$
   n. $\frac{a-b}{xa-xb-a+b}$
   o. $\frac{(x+y)^2-9a^2}{2x+2y-6a}$
   p. $\frac{8x^3-y^3}{4x^2-y^2}$

2. Write a rational expression with denominator $6b$ that is equivalent to

   a. $\frac{a}{b}$.
   b. one-half of $\frac{a}{b}$.
   c. $\frac{1}{3}$.

3. Remember that algebra is just a symbolic method for performing arithmetic.

   a. Simplify the following rational expression: $\frac{(x^2y)^2(xy)^3z^2}{(xy^2)^2yz}$.
   b. Simplify the following rational expression without using a calculator: $\frac{12^2 \cdot 6^3 \cdot 5^2}{18^2 \cdot 15}$.
   c. How are the calculations in parts (a) and (b) similar? How are they different? Which expression was easier to simplify?

This page intentionally left blank

Lesson 23: Comparing Rational Expressions

Classwork

Opening Exercise

Use the slips of paper you have been given to create visual arguments for whether $\frac{1}{3}$ or $\frac{3}{8}$ is larger.

Exercises

We will start by working with positive integers. Let m and n be positive integers. Work through the following exercises with a partner.

1. Fill out the following table.

n	$\frac{1}{n}$
1	
2	
3	
4	
5	
6	

2. Do you expect $\frac{1}{n}$ to be larger or smaller than $\frac{1}{n+1}$? Do you expect $\frac{1}{n}$ to be larger or smaller than $\frac{1}{n+2}$? Explain why.

3. Compare the rational expressions $\frac{1}{n}$, $\frac{1}{n+1}$, and $\frac{1}{n+2}$ for $n = 1, 2$, and 3. Do your results support your conjecture from Exercise 2? Revise your conjecture if necessary.

4. From your work in Exercises 1 and 2, generalize how $\frac{1}{n}$ compares to $\frac{1}{n+m}$, where m and n are positive integers.

5. Will your conjecture change or stay the same if the numerator is 2 instead of 1? Make a conjecture about what happens when the numerator is held constant, but the denominator increases for positive numbers.

Example

x	$\frac{x+1}{x}$	$\frac{x+2}{x+1}$
0.5		
1		
1.5		
2		
5		
10		
100		

Lesson Summary

To compare two rational expressions, find equivalent rational expression with the same denominator. Then we can compare the numerators for values of the variable that do not cause the rational expression to change from positive to negative or vice versa.

We may also use numerical and graphical analysis to help understand the relative sizes of expressions.

Problem Set

1. For parts (a)–(d), rewrite each rational expression as an equivalent rational expression so that all expressions have a common denominator.
   a. $\frac{3}{5}, \frac{9}{10}, \frac{7}{15}, \frac{7}{21}$
   b. $\frac{m}{sd}, \frac{s}{dm}, \frac{d}{ms}$
   c. $\frac{1}{(2-x)^2}, \frac{3}{(2x-5)(x-2)}$
   d. $\frac{3}{x-x^2}, \frac{5}{x}, \frac{2x+2}{2x^2-2}$

2. If x is a positive number, for which values of x is $x < \frac{1}{x}$?

3. Can we determine if $\frac{y}{y-1} > \frac{y+1}{y}$ for all values $y > 1$? Provide evidence to support your answer.

4. For positive x, determine when the following rational expressions have negative denominators.
   a. $\frac{3}{5}$
   b. $\frac{x}{5-2x}$
   c. $\frac{x+3}{x^2+4x+8}$
   d. $\frac{3x^2}{(x-5)(x+3)(2x+3)}$

5. Consider the rational expressions $\frac{x}{x-2}$ and $\frac{x}{x-4}$.

   a. Evaluate each expression for $x = 6$.

   b. Evaluate each expression for $x = 3$.

   c. Can you conclude that $\frac{x}{x-2} < \frac{x}{x-4}$ for all positive values of x? Explain how you know.

   d. Extension: Raphael claims that the calculation below shows that $\frac{x}{x-2} < \frac{x}{x-4}$ for all values of x, where $x \neq 2$ and $x \neq 4$. Where is the error in the calculation?

   Starting with the rational expressions $\frac{x}{x-2}$ and $\frac{x}{x-4}$, we need to first find equivalent rational expressions with a common denominator. The common denominator we will use is $(x-4)(x-2)$. We then have

$$\frac{x}{x-2} = \frac{x(x-4)}{(x-4)(x-2)}$$
$$\frac{x}{x-4} = \frac{x(x-2)}{(x-4)(x-2)}.$$

   Since $x^2 - 4x < x^2 - 2x$ for $x > 0$, we can divide each expression by $(x-4)(x-2)$. We then have $\frac{x(x-4)}{(x-4)(x-2)} < \frac{x(x-2)}{(x-4)(x-2)}$, and we can conclude that $\frac{x}{x-2} < \frac{x}{x-4}$ for all positive values of x.

6. Consider the populations of two cities within the same state where the large city's population is P, and the small city's population is Q. For each of the following pairs, state which of the expressions has a larger value. Explain your reasoning in the context of the populations.

   a. $P + Q$ and P

   b. $\frac{P}{P+Q}$ and $\frac{Q}{P+Q}$

   c. $2Q$ and $P + Q$

   d. $\frac{P}{Q}$ and $\frac{Q}{P}$

   e. $\frac{P}{P+Q}$ and $\frac{1}{2}$

   f. $\frac{P+Q}{P}$ and $P - Q$

   g. $\frac{P+Q}{2}$ and $\frac{P+Q}{Q}$

   h. $\frac{1}{P}$ and $\frac{1}{Q}$

This page intentionally left blank

Lesson 24: Multiplying and Dividing Rational Expressions

Classwork

If a, b, c, and d are rational expressions with $b \neq 0$, $d \neq 0$, then

$$\frac{a}{b} \cdot \frac{c}{d} = \frac{ac}{bd}.$$

Example 1

Make a conjecture about the product $\frac{x^3}{4y} \cdot \frac{y^2}{x}$. What will it be? Explain your conjecture, and give evidence that it is correct.

Example 2

Find the following product:

$$\left(\frac{3x-6}{2x+6}\right) \cdot \left(\frac{5x+15}{4x+8}\right).$$

Exercises 1–3

1. Summarize what you have learned so far with your neighbor.

2. Find the following product and reduce to lowest terms: $\left(\frac{2x+6}{x^2+x-6}\right) \cdot \left(\frac{x^2-4}{2x}\right)$.

3. Find the following product and reduce to lowest terms: $\left(\frac{4n-12}{3m+6}\right)^{-2} \cdot \left(\frac{n^2-2n-3}{m^2+4m+4}\right)$.

If a, b, c, and d are rational expressions with $b \neq 0$, $c \neq 0$, and $d \neq 0$, then

$$\frac{a}{b} \div \frac{c}{d} = \frac{a}{b} \cdot \frac{d}{c} = \frac{ad}{bc}.$$

Example 3

Find the quotient and reduce to lowest terms: $\frac{x^2-4}{3x} \div \frac{x-2}{2x}$.

Exercises 4–5

4. Find the quotient and reduce to lowest terms: $\frac{x^2-5x+6}{x+4} \div \frac{x^2-9}{x^2+5x+4}$.

5. Simplify the rational expression.

$$\frac{\left(\frac{x+2}{x^2-2x-3}\right)}{\left(\frac{x^2-x-6}{x^2+6x+5}\right)}$$

Lesson Summary

In this lesson, we extended multiplication and division of rational numbers to multiplication and division of rational expressions.

- To multiply two rational expressions, multiply the numerators together and multiply the denominators together, and then reduce to lowest terms.
- To divide one rational expression by another, multiply the first by the multiplicative inverse of the second, and reduce to lowest terms.
- To simplify a complex fraction, apply the process for dividing one rational expression by another.

Problem Set

1. Perform the following operations:

   a. Multiply $\frac{1}{3}(x-2)$ by 9.

   b. Divide $\frac{1}{4}(x-8)$ by $\frac{1}{12}$.

   c. Multiply $\frac{1}{4}\left(\frac{1}{3}x+2\right)$ by 12.

   d. Divide $\frac{1}{3}\left(\frac{2}{5}x-\frac{1}{5}\right)$ by $\frac{1}{15}$.

   e. Multiply $\frac{2}{3}\left(2x+\frac{2}{3}\right)$ by $\frac{9}{4}$.

   f. Multiply $0.03(4-x)$ by 100.

2. Write each rational expression as an equivalent rational expression in lowest terms.

   a. $\left(\frac{a^3b^2}{c^2d^2}\cdot\frac{c}{ab}\right)\div\frac{a}{c^2d^3}$

   b. $\frac{a^2+6a+9}{a^2-9}\cdot\frac{3a-9}{a+3}$

   c. $\frac{6x}{4x-16}\div\frac{4x}{x^2-16}$

   d. $\frac{3x^2-6x}{3x+1}\cdot\frac{x+3x^2}{x^2-4x+4}$

   e. $\frac{2x^2-10x+12}{x^2-4}\cdot\frac{2+x}{3-x}$

   f. $\frac{a-2b}{a+2b}\div(4b^2-a^2)$

   g. $\frac{d+c}{c^2+d^2}\div\frac{c^2-d^2}{d^2-dc}$

   h. $\frac{12a^2-7ab+b^2}{9a^2-b^2}\div\frac{16a^2-b^2}{3ab+b^2}$

   i. $\left(\frac{x-3}{x^2-4}\right)^{-1}\cdot\left(\frac{x^2-x-6}{x-2}\right)$

   j. $\left(\frac{x-2}{x^2+1}\right)^{-3}\div\left(\frac{x^2-4x+4}{x^2-2x-3}\right)$

   k. $\frac{6x^2-11x-10}{6x^2-5x-6}\cdot\frac{6-4x}{25-20x+4x^2}$

   l. $\frac{3x^3-3a^2x}{x^2-2ax+a^2}\cdot\frac{a-x}{a^3x+a^2x^2}$

3. Write each rational expression as an equivalent rational expression in lowest terms.

   a. $\dfrac{\left(\frac{4a}{6b^2}\right)}{\left(\frac{20a^3}{12b}\right)}$

   b. $\dfrac{\left(\frac{x-2}{x^2-1}\right)}{\left(\frac{x^2-4}{x-6}\right)}$

   c. $\dfrac{\left(\frac{x^2+2x-3}{x^2+3x-4}\right)}{\left(\frac{x^2+x-6}{x+4}\right)}$

4. Suppose that $x = \frac{t^2+3t-4}{3t^2-3}$ and $y = \frac{t^2+2t-8}{2t^2-2t-4}$, for $t \neq 1$, $t \neq -1$, $t \neq 2$, and $t \neq -4$. Show that the value of x^2y^{-2} does not depend on the value of t.

5. Determine which of the following numbers is larger without using a calculator, $\frac{15^{16}}{16^{15}}$ or $\frac{20^{24}}{24^{20}}$. (Hint: We can compare two positive quantities a and b by computing the quotient $\frac{a}{b}$. If $\frac{a}{b} > 1$, then $a > b$. Likewise, if $0 < \frac{a}{b} < 1$, then $a < b$.)

Extension:

6. One of two numbers can be represented by the rational expression $\frac{x-2}{x}$, where $x \neq 0$ and $x \neq 2$.

   a. Find a representation of the second number if the product of the two numbers is 1.

   b. Find a representation of the second number if the product of the two numbers is 0.

This page intentionally left blank

Lesson 25: Adding and Subtracting Rational Expressions

Classwork

Exercises 1–4

1. Calculate the following sum: $\frac{3}{10} + \frac{6}{10}$.

2. $\frac{3}{20} - \frac{4}{15}$

3. $\frac{\pi}{4} + \frac{\sqrt{2}}{5}$

4. $\frac{a}{m} + \frac{b}{2m} - \frac{c}{m}$

Example 1

Perform the indicated operations below and simplify.

a. $\frac{a+b}{4}+\frac{2a-b}{5}$

b. $\frac{4}{3x}-\frac{3}{5x^2}$

c. $\frac{3}{2x^2+2x}+\frac{5}{x^2-3x-4}$

Exercises 5–8

Perform the indicated operations for each problem below.

5. $\frac{5}{x-2} + \frac{3x}{4x-8}$

6. $\frac{7m}{m-3} + \frac{5m}{3-m}$

7. $\frac{b^2}{b^2-2bc+c^2} - \frac{b}{b-c}$

8. $\frac{x}{x^2-1} - \frac{2x}{x^2+x-2}$

Example 2

Simplify the following expression.

$$\frac{\frac{b^2+b-1}{2b-1}-1}{4-\frac{8}{(b+1)}}$$

Lesson Summary

In this lesson, we extended addition and subtraction of rational numbers to addition and subtraction of rational expressions. The process for adding or subtracting rational expressions can be summarized as follows:

- Find a common multiple of the denominators to use as a common denominator.
- Find equivalent rational expressions for each expression using the common denominator.
- Add or subtract the numerators as indicated and simplify if needed.

Problem Set

1. Write each sum or difference as a single rational expression.

   a. $\frac{7}{8} - \frac{\sqrt{3}}{5}$

   b. $\frac{\sqrt{5}}{10} + \frac{\sqrt{2}}{6} + 2$

   c. $\frac{4}{x} + \frac{3}{2x}$

2. Write as a single rational expression.

   a. $\frac{1}{x} - \frac{1}{x-1}$

   b. $\frac{3x}{2y} - \frac{5x}{6y} + \frac{x}{3y}$

   c. $\frac{a-b}{a^2} + \frac{1}{a}$

   d. $\frac{1}{p-2} - \frac{1}{p+2}$

   e. $\frac{1}{p-2} + \frac{1}{2-p}$

   f. $\frac{1}{b+1} - \frac{b}{1+b}$

   g. $1 - \frac{1}{1+p}$

   h. $\frac{p+q}{p-q} - 2$

   i. $\frac{r}{s-r} + \frac{s}{r+s}$

   j. $\frac{3}{x-4} + \frac{2}{4-x}$

   k. $\frac{3n}{n-2} + \frac{3}{2-n}$

   l. $\frac{8x}{3y-2x} + \frac{12y}{2x-3y}$

   m. $\frac{1}{2m-4n} - \frac{1}{2m+4n} - \frac{m}{m^2-4n^2}$

   n. $\frac{1}{(2a-b)(a-c)} + \frac{1}{(b-c)(b-2a)}$

   o. $\frac{b^2+1}{b^2-4} + \frac{1}{b+2} + \frac{1}{b-2}$

3. Write each rational expression as an equivalent rational expression in lowest terms.

   a. $\frac{\frac{1}{a} - \frac{1}{2a}}{\frac{4}{a}}$

   b. $\frac{\frac{5x}{2} + 1}{\frac{5x}{4} - \frac{1}{5x}}$

   c. $\frac{1 + \frac{4x+3}{x^2+1}}{1 - \frac{x+7}{x^2+1}}$

Extension:

4. Suppose that $x \neq 0$ and $y \neq 0$. We know from our work in this section that $\frac{1}{x} \cdot \frac{1}{y}$ is equivalent to $\frac{1}{xy}$. Is it also true that $\frac{1}{x} + \frac{1}{y}$ is equivalent to $\frac{1}{x+y}$? Provide evidence to support your answer.

5. Suppose that $x = \frac{2t}{1+t^2}$ and $y = \frac{1-t^2}{1+t^2}$. Show that the value of $x^2 + y^2$ does not depend on the value of t.

6. Show that for any real numbers a and b, and any integers x and y so that $x \neq 0$, $y \neq 0$, $x \neq y$, and $x \neq -y$,

$$\left(\frac{y}{x} - \frac{x}{y}\right)\left(\frac{ax+by}{x+y} - \frac{ax-by}{x-y}\right) = 2(a-b).$$

7. Suppose that n is a positive integer.

   a. Rewrite the product in the form $\frac{P}{Q}$ for polynomials P and Q: $\left(1+\frac{1}{n}\right)\left(1+\frac{1}{n+1}\right)$.

   b. Rewrite the product in the form $\frac{P}{Q}$ for polynomials P and Q: $\left(1+\frac{1}{n}\right)\left(1+\frac{1}{n+1}\right)\left(1+\frac{1}{n+2}\right)$.

   c. Rewrite the product in the form $\frac{P}{Q}$ for polynomials P and Q: $\left(1+\frac{1}{n}\right)\left(1+\frac{1}{n+1}\right)\left(1+\frac{1}{n+2}\right)\left(1+\frac{1}{n+3}\right)$.

   d. If this pattern continues, what is the product of n of these factors?

Lesson 26: Solving Rational Equations

Classwork

Exercises 1–2

Solve the following equations for x, and give evidence that your solutions are correct.

1. $\frac{x}{2} + \frac{1}{3} = \frac{5}{6}$

2. $\frac{2x}{9} + \frac{5}{9} = \frac{8}{9}$

Example

Solve the following equation: $\frac{x+3}{12} = \frac{5}{6}$.

Exercises 3–7

3. Solve the following equation: $\frac{3}{x} = \frac{8}{x-2}$.

4. Solve the following equation for a: $\frac{1}{a+2} + \frac{1}{a-2} = \frac{4}{a^2-4}$.

5. Solve the following equation. Remember to check for extraneous solutions.

$$\frac{4}{3x} + \frac{5}{4} = \frac{3}{x}$$

6. Solve the following equation. Remember to check for extraneous solutions.

$$\frac{7}{b+3}+\frac{5}{b-3}=\frac{10b-2}{b^2-9}$$

7. Solve the following equation. Remember to check for extraneous solutions.

$$\frac{1}{x-6}+\frac{x}{x-2}=\frac{4}{x^2-8x+12}$$

Lesson Summary

In this lesson, we applied what we have learned in the past two lessons about addition, subtraction, multiplication, and division of rational expressions to solve rational equations. An extraneous solution is a solution to a transformed equation that is not a solution to the original equation. For rational functions, extraneous solutions come from the excluded values of the variable.

Rational equations can be solved one of two ways:

1. Write each side of the equation as an equivalent rational expression with the same denominator and equate the numerators. Solve the resulting polynomial equation, and check for extraneous solutions.
2. Multiply both sides of the equation by an expression that is the common denominator of all terms in the equation. Solve the resulting polynomial equation, and check for extraneous solutions.

Problem Set

1. Solve the following equations, and check for extraneous solutions.

   a. $\frac{x-8}{x-4} = 2$

   b. $\frac{4x-8}{x-2} = 4$

   c. $\frac{x-4}{x-3} = 1$

   d. $\frac{4x-8}{x-2} = 3$

   e. $\frac{1}{2a} - \frac{2}{2a-3} = 0$

   f. $\frac{3}{2x+1} = \frac{5}{4x+3}$

   g. $\frac{4}{x-5} - \frac{2}{5+x} = \frac{2}{x}$

   h. $\frac{y+2}{3y-2} + \frac{y}{y-1} = \frac{2}{3}$

   i. $\frac{3}{x+1} - \frac{2}{1-x} = 1$

   j. $\frac{4}{x-1} + \frac{3}{x} - 3 = 0$

   k. $\frac{x+1}{x+3} - \frac{x-5}{x+2} = \frac{17}{6}$

   l. $\frac{x+7}{4} - \frac{x+1}{2} = \frac{5-x}{3x-14}$

   m. $\frac{b^2-b-6}{b^2} - \frac{2b+12}{b} = \frac{b-39}{2b}$

   n. $\frac{1}{p(p-4)} + 1 = \frac{p-6}{p}$

   o. $\frac{1}{h+3} = \frac{h+4}{h-2} + \frac{6}{h-2}$

   p. $\frac{m+5}{m^2+m} = \frac{1}{m^2+m} - \frac{m-6}{m+1}$

2. Create and solve a rational equation that has 0 as an extraneous solution.

3. Create and solve a rational equation that has 2 as an extraneous solution.

Extension:

4. Two lengths a and b, where $a > b$, are in *golden ratio* if the ratio of $a + b$ is to a is the same as a is to b. Symbolically, this is expressed as $\frac{a}{b} = \frac{a+b}{a}$. We denote this common ratio by the Greek letter *phi (pronounced "fee")* with symbol φ, so that if a and b are in common ratio, then $\varphi = \frac{a}{b} = \frac{a+b}{a}$. By setting $b = 1$, we find that $\varphi = a$ and φ is the positive number that satisfies the equation $\varphi = \frac{\varphi+1}{\varphi}$. Solve this equation to find the numerical value for φ.

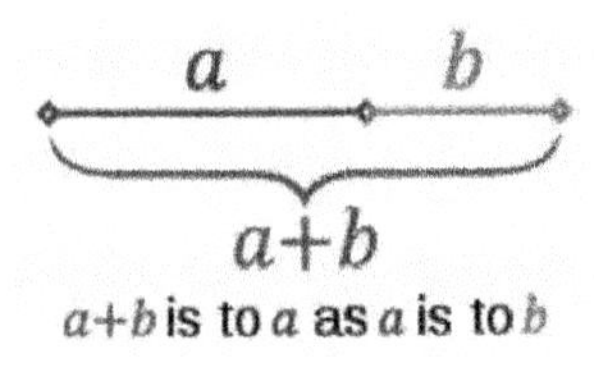

5. Remember that if we use x to represent an integer, then the next integer can be represented by $x + 1$.
   a. Does there exist a pair of consecutive integers whose reciprocals sum to $\frac{5}{6}$? Explain how you know.
   b. Does there exist a pair of consecutive integers whose reciprocals sum to $\frac{3}{4}$? Explain how you know.
   c. Does there exist a pair of consecutive *even* integers whose reciprocals sum to $\frac{3}{4}$? Explain how you know.
   d. Does there exist a pair of consecutive *even* integers whose reciprocals sum to $\frac{5}{6}$? Explain how you know.

This page intentionally left blank

Lesson 27: Word Problems Leading to Rational Equations

Classwork

Exercise 1

1. Anne and Maria play tennis almost every weekend. So far, Anne has won 12 out of 20 matches.
   a. How many matches will Anne have to win in a row to improve her winning percentage to 75%?

   b. How many matches will Anne have to win in a row to improve her winning percentage to 90%?

   c. Can Anne reach a winning percentage of 100%?

d. After Anne has reached a winning percentage of 90% by winning consecutive matches as in part (b), how many matches can she now lose in a row to have a winning percentage of 50%?

Example

Working together, it takes Sam, Jenna, and Francisco two hours to paint one room. When Sam works alone, he can paint one room in 6 hours. When Jenna works alone, she can paint one room in 4 hours. Determine how long it would take Francisco to paint one room on his own.

Exercises 2–4

2. Melissa walks 3 miles to the house of a friend and returns home on a bike. She averages 4 miles per hour faster when cycling than when walking, and the total time for both trips is two hours. Find her walking speed.

3. You have 10 liters of a juice blend that is 60% juice.

   a. How many liters of pure juice need to be added in order to make a blend that is 75% juice?

   b. How many liters of pure juice need to be added in order to make a blend that is 90% juice?

   c. Write a rational equation that relates the desired percentage p to the amount A of pure juice that needs to be added to make a blend that is $p\%$ juice, where $0 < p < 100$. What is a reasonable restriction on the set of possible values of p? Explain your answer.

   d. Suppose that you have added 15 liters of juice to the original 10 liters. What is the percentage of juice in this blend?

e. Solve your equation in part (c) for the amount A. Are there any excluded values of the variable p? Does this make sense in the context of the problem?

4. You have a solution containing 10% acid and a solution containing 30% acid.

a. How much of the 30% solution must you add to 1 liter of the 10% solution to create a mixture that is 22% acid?

b. Write a rational equation that relates the desired percentage p to the amount A of 30% acid solution that needs to be added to 1 liter of 10% acid solution to make a blend that is $p\%$ acid, where $0 < p < 100$. What is a reasonable restriction on the set of possible values of p? Explain your answer.

c. Solve your equation in part (b) for A. Are there any excluded values of p? Does this make sense in the context of the problem?

d. If you have added some 30% acid solution to 1 liter of 10% acid solution to make a 26% acid solution, how much of the stronger acid did you add?

Problem Set

1. If two inlet pipes can fill a pool in one hour and 30 minutes, and one pipe can fill the pool in two hours and 30 minutes on its own, how long would the other pipe take to fill the pool on its own?

2. If one inlet pipe can fill the pool in 2 hours with the outlet drain closed, and the same inlet pipe can fill the pool in 2.5 hours with the drain open, how long does it take the drain to empty the pool if there is no water entering the pool?

3. It takes 36 minutes less time to travel 120 miles by car at night than by day because the lack of traffic allows the average speed at night to be 10 miles per hour faster than in the daytime. Find the average speed in the daytime.

4. The difference in the average speed of two trains is 16 miles per hour. The slower train takes 2 hours longer to travel 170 miles than the faster train takes to travel 150 miles. Find the speed of the slower train.

5. A school library spends \$80 a month on magazines. The average price for magazines bought in January was 70 cents more than the average price in December. Because of the price increase, the school library was forced to subscribe to 7 fewer magazines. How many magazines did the school library subscribe to in December?

6. An investor bought a number of shares of stock for \$1,600. After the price dropped by \$10 per share, the investor sold all but 4 of her shares for \$1,120. How many shares did she originally buy?

7. Newton's law of universal gravitation, $F = \frac{Gm_1m_2}{r^2}$, measures the force of gravity between two masses m_1 and m_2, where r is the distance between the centers of the masses, and G is the universal gravitational constant. Solve this equation for G.

8. Suppose that $t = \frac{x+y}{1-xy}$.

   a. Show that when $x = \frac{1}{a}$ and $y = \frac{2a-1}{a+2}$, the value of t does not depend on the value of a.

   b. For which values of a do these relationships have no meaning?

9. Consider the rational equation $\frac{1}{R} = \frac{1}{x} + \frac{1}{y}$.

   a. Find the value of R when $x = \frac{2}{5}$ and $y = \frac{3}{4}$.

   b. Solve this equation for R, and write R as a single rational expression in lowest terms.

10. Consider an ecosystem of rabbits in a park that starts with 10 rabbits and can sustain up to 60 rabbits. An equation that roughly models this scenario is

$$P = \frac{60}{1 + \frac{5}{t+1}},$$

where P represents the rabbit population in year t of the study.

   a. What is the rabbit population in year 10? Round your answer to the nearest whole rabbit.
   b. Solve this equation for t. Describe what this equation represents in the context of this problem.
   c. At what time does the population reach 50 rabbits?

Extension:

11. Suppose that Huck Finn can paint a fence in 5 hours. If Tom Sawyer helps him pain the fence, they can do it in 3 hours. How long would it take for Tom to paint the fence by himself?

12. Huck Finn can paint a fence in 5 hours. After some practice, Tom Sawyer can now paint the fence in 6 hours.
   a. How long would it take Huck and Tom to paint the fence together?
   b. Tom demands a half-hour break while Huck continues to paint, and they finish the job together. How long does it take them to paint the fence?
   c. Suppose that they have to finish the fence in $3\frac{1}{2}$ hours. What's the longest break that Tom can take?

This page intentionally left blank

Lesson 28: A Focus on Square Roots

Classwork

Exercises 1–4

For Exercises 1–4, describe each step taken to solve the equation. Then, check the solution to see if it is valid. If it is not a valid solution, explain why.

1. $\sqrt{x} - 6 = 4$

   $\sqrt{x} = 10$

   $x = 100$

2. $\sqrt[3]{x} - 6 = 4$

   $\sqrt[3]{x} = 10$

   $x = 1000$

3. $\sqrt{x} + 6 = 4$

4. $\sqrt[3]{x} + 6 = 4$

Example 1

Solve the following radical equation. Be sure to check your solutions.

$$\sqrt{3x+5} - 2 = -1$$

Exercises 5–15

Solve each radical equation. Be sure to check your solutions.

5. $\sqrt{2x-3} = 11$

6. $\sqrt[3]{6-x} = -3$

7. $\sqrt{x+5} - 9 = -12$

8. $\sqrt{4x-7} = \sqrt{3x+9}$

9. $-12\sqrt{x-6} = 18$

10. $3\sqrt[3]{x+2} = 12$

11. $\sqrt{x^2-5} = 2$

12. $\sqrt{x^2+8x} = 3$

Compute each product, and combine like terms.

13. $(\sqrt{x}+2)(\sqrt{x}-2)$

14. $(\sqrt{x}+4)(\sqrt{x}+4)$

15. $(\sqrt{x-5})(\sqrt{x-5})$

Example 2

Rationalize the denominator in each expression. That is, rewrite the expression so that there is a rational expression in the denominator.

a. $\frac{x-9}{\sqrt{x-9}}$

b. $\frac{x-9}{\sqrt{x}+3}$

Exercises 16–18

16. Rewrite $\frac{1}{\sqrt{x}-5}$ in an equivalent form with a rational expression in the denominator.

17. Solve the radical equation $\frac{3}{\sqrt{x+3}} = 1$. Be sure to check for extraneous solutions.

18. Without solving the radical equation $\sqrt{x+5} + 9 = 0$, how could you tell that it has no real solution?

Problem Set

1.
   a. If $\sqrt{x} = 9$, then what is the value of x?
   b. If $x^2 = 9$, then what is the value of x?
   c. Is there a value of x such that $\sqrt{x+5} = 0$? If yes, what is the value? If no, explain why not.
   d. Is there a value of x such that $\sqrt{x} + 5 = 0$? If yes, what is the value? If no, explain why not.

2.
   a. Is the statement $\sqrt{x^2} = x$ true for all x-values? Explain.
   b. Is the statement $\sqrt[3]{x^3} = x$ true for all x-values? Explain.

Rationalize the denominator in each expression.

3. $\frac{4-x}{2+\sqrt{x}}$

4. $\frac{2}{\sqrt{x-12}}$

5. $\frac{1}{\sqrt{x+3}-\sqrt{x}}$

Solve each equation, and check the solutions.

6. $\sqrt{x+6} = 3$

7. $2\sqrt{x+3} = 6$

8. $\sqrt{x+3} + 6 = 3$

9. $\sqrt{x+3} - 6 = 3$

10. $16 = 8 + \sqrt{x}$

11. $\sqrt{3x-5} = 7$

12. $\sqrt{2x-3} = \sqrt{10-x}$

13. $3\sqrt{x+2} + \sqrt{x-4} = 0$

14. $\frac{\sqrt{x+9}}{4} = 3$

15. $\frac{12}{\sqrt{x+9}} = 3$

16. $\sqrt{x^2+9} = 5$

17. $\sqrt{x^2-6x} = 4$

18. $\frac{5}{\sqrt{x-2}} = 5$

19. $\frac{5}{\sqrt{x}-2} = 5$

20. $\sqrt[3]{5x-3} + 8 = 6$

21. $\sqrt[3]{9-x} = 6$

22. Consider the inequality $\sqrt{x^2+4x} > 0$. Determine whether each x-value is a solution to the inequality.

   a. $x = -10$
   b. $x = -4$
   c. $x = 10$
   d. $x = 4$

23. Show that $\frac{a-b}{\sqrt{a}-\sqrt{b}} = \sqrt{a} + \sqrt{b}$ for all values of a and b such that $a > 0$ and $b > 0$ and $a \neq b$.

24. Without actually solving the equation, explain why the equation $\sqrt{x+1} + 2 = 0$ has no solution.

Lesson 29: Solving Radical Equations

Classwork

Example 1

Solve the equation $6 = x + \sqrt{x}$.

Exercises 1–4

Solve.

1. $3x = 1 + 2\sqrt{x}$

2. $3 = 4\sqrt{x} - x$

3. $\sqrt{x+5} = x - 1$

4. $\sqrt{3x+7} + 2\sqrt{x-8} = 0$

Example 2

Solve the equation $\sqrt{x} + \sqrt{x+3} = 3$.

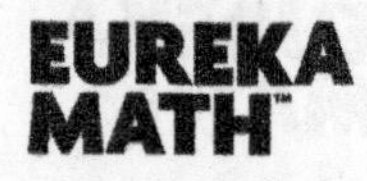

Exercises 5–6

Solve the following equations.

5. $\sqrt{x-3}+\sqrt{x+5}=4$

6. $3+\sqrt{x}=\sqrt{x+81}$

Lesson Summary

If $a = b$ and n is an integer, then $a^n = b^n$. However, the converse is not necessarily true. The statement $a^n = b^n$ does not imply that $a = b$. Therefore, it is necessary to check for extraneous solutions when both sides of an equation are raised to an exponent.

Problem Set

Solve.

1. $\sqrt{2x-5} - \sqrt{x+6} = 0$
2. $\sqrt{2x-5} + \sqrt{x+6} = 0$
3. $\sqrt{x-5} - \sqrt{x+6} = 2$
4. $\sqrt{2x-5} - \sqrt{x+6} = 2$
5. $\sqrt{x+4} = 3 - \sqrt{x}$
6. $\sqrt{x+4} = 3 + \sqrt{x}$
7. $\sqrt{x+3} = \sqrt{5x+6} - 3$
8. $\sqrt{2x+1} = x - 1$
9. $\sqrt{x+12} + \sqrt{x} = 6$
10. $2\sqrt{x} = 1 - \sqrt{4x-1}$
11. $2x = \sqrt{4x-1}$
12. $\sqrt{4x-1} = 2 - 2x$
13. $x + 2 = 4\sqrt{x-2}$
14. $\sqrt{2x-8} + \sqrt{3x-12} = 0$
15. $x = 2\sqrt{x-4} + 4$
16. $x - 2 = \sqrt{9x-36}$
17. Consider the right triangle ABC shown to the right, with $AB = 8$ and $BC = x$.
    a. Write an expression for the length of the hypotenuse in terms of x.
    b. Find the value of x for which $AC - AB = 9$.

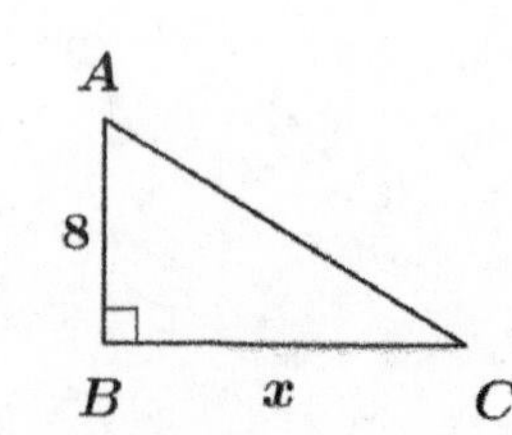

18. Consider the triangle ABC shown to the right where $AD = DC$, and $\overline{BD}$ is the altitude of the triangle.
    a. If the length of $\overline{BD}$ is x cm, and the length of $\overline{AC}$ is 18 cm, write an expression for the lengths of $\overline{AB}$ and $\overline{BC}$ in terms of x.
    b. Write an expression for the perimeter of $\triangle ABC$ in terms of x.
    c. Find the value of x for which the perimeter of $\triangle ABC$ is equal to 38 cm.

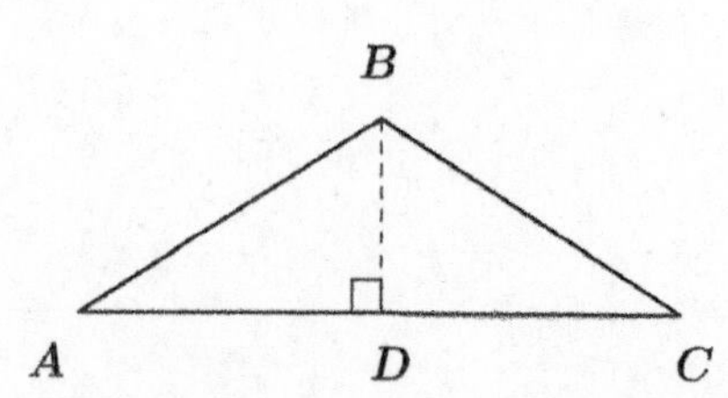

Lesson 30: Linear Systems in Three Variables

Classwork

Exercises 1–3

Determine the value of x and y in the following systems of equations.

1. $2x + 3y = 7$
   $2x + y = 3$

2. $5x - 2y = 4$
   $-2x + y = 2$

3. A scientist wants to create 120 ml of a solution that is 30% acidic. To create this solution, she has access to a 20% solution and a 45% solution. How many milliliters of each solution should she combine to create the 30% solution?

Example

Determine the values for x, y, and z in the following system:

$$2x + 3y - z = 5$$
$$4x - y - z = -1$$
$$x + 4y + z = 12.$$

Exercises 4–5

4. Given the system below, determine the values of r, s, and u that satisfy all three equations.

$$r + 2x - u = 8$$
$$s + u = 4$$
$$r - s - u = 2$$

5. Find the equation of the form $y = ax^2 + bx + c$ whose graph passes through the points $(1, 6)$, $(3, 20)$, and $(-2, 15)$.

Problem Set

Solve the following systems of equations.

1. $x + y = 3$
   $y + z = 6$
   $x + z = 5$

2. $r = 2(s - t)$
   $2t = 3(s - r)$
   $r + t = 2s - 3$

3. $2a + 4b + c = 5$
   $a - 4b = -6$
   $2b + c = 7$

4. $2x + y - z = -5$
   $4x - 2y + z = 10$
   $2x + 3y + 2z = 3$

5. $r + 3s + t = 3$
   $2r - 3s + 2t = 3$
   $-r + 3s - 3t = 1$

6. $x - y = 1$
   $2y + z = -4$
   $x - 2z = -6$

7. $x = 3(y - z)$
   $y = 5(z - x)$
   $x + y = z + 4$

8. $p + q + 3r = 4$
   $2q + 3r = 7$
   $p - q - r = -2$

9. $\frac{1}{x} + \frac{1}{y} + \frac{1}{z} = 5$
   $\frac{1}{x} + \frac{1}{y} = 2$
   $\frac{1}{x} - \frac{1}{z} = -2$

10. $\frac{1}{a} + \frac{1}{b} + \frac{1}{c} = 6$
    $\frac{1}{b} + \frac{1}{c} = 5$
    $\frac{1}{a} - \frac{1}{b} = -1$

11. Find the equation of the form $y = ax^2 + bx + c$ whose graph passes through the points $(1, -1)$, $(3, 23)$, and $(-1, 7)$.

12. Show that for any number t, the values $x = t + 2$, $y = 1 - t$, and $z = t + 1$ are solutions to the system of equations below.

$$x + y = 3$$
$$y + z = 2$$

(In this situation, we say that t *parameterizes the solution set of the system.*)

13. Some rational expressions can be written as the sum of two or more rational expressions whose denominators are the factors of its denominator (called a *partial fraction decomposition*). Find the partial fraction decomposition for $\frac{1}{n(n+1)}$ by finding the value of A that makes the equation below true for all n except 0 and -1.

$$\frac{1}{n(n+1)} = \frac{A}{n} - \frac{1}{n+1}$$

14. A chemist needs to make 40 ml of a 15% acid solution. He has a 5% acid solution and a 30% acid solution on hand. If he uses the 5% and 30% solutions to create the 15% solution, how many ml of each does he need?

15. An airplane makes a 400-mile trip against a head wind in 4 hours. The return trip takes 2.5 hours, the wind now being a tall wind. If the plane maintains a constant speed with respect to still air, and the speed of the wind is also constant and does not vary, find the still-air speed of the plane and the speed of the wind.

16. A restaurant owner estimates that she needs the same number of pennies as nickels and the same number of dimes as pennies and nickels together. How should she divide $\$26$ between pennies, nickels, and dimes?

Lesson 31: Systems of Equations

Classwork

Exploratory Challenge 1

a. Sketch the lines given by $x + y = 6$ and $-3x + y = 2$ on the same set of axes to solve the system graphically. Then solve the system of equations algebraically to verify your graphical solution.

b. Suppose the second line is replaced by the line with equation $x + y = 2$. Plot the two lines on the same set of axes, and solve the pair of equations algebraically to verify your graphical solution.

c. Suppose the second line is replaced by the line with equation $2x = 12 - 2y$. Plot the lines on the same set of axes, and solve the pair of equations algebraically to verify your graphical solution.

d. We have seen that a pair of lines can intersect in 1, 0, or an infinite number of points. Are there any other possibilities?

Exploratory Challenge 2

a. Suppose that instead of equations for a pair of lines, you were given an equation for a circle and an equation for a line. What possibilities are there for the two figures to intersect? Sketch a graph for each possibility.

b. Graph the parabola with equation $y = x^2$. What possibilities are there for a line to intersect the parabola? Sketch each possibility.

c. Sketch the circle given by $x^2 + y^2 = 1$ and the line given by $y = 2x + 2$ on the same set of axes. One solution to the pair of equations is easily identifiable from the sketch. What is it?

d. Substitute $y = 2x + 2$ into the equation $x^2 + y^2 = 1$, and solve the resulting equation for x.

e. What does your answer to part (d) tell you about the intersections of the circle and the line from part (c)?

Exercises

1. Draw a graph of the circle with equation $x^2 + y^2 = 9$.

   a. What are the solutions to the system of circle and line when the circle is given by $x^2 + y^2 = 9$, and the line is given by $y = 2$?

b. What happens when the line is given by $y = 3$?

c. What happens when the line is given by $y = 4$?

2. By solving the equations as a system, find the points common to the line with equation $x - y = 6$ and the circle with equation $x^2 + y^2 = 26$. Graph the line and the circle to show those points.

3. Graph the line given by $5x + 6y = 12$ and the circle given by $x^2 + y^2 = 1$. Find all solutions to the system of equations.

4. Graph the line given by $3x + 4y = 25$ and the circle given by $x^2 + y^2 = 25$. Find all solutions to the system of equations. Verify your result both algebraically and graphically.

5. Graph the line given by $2x + y = 1$ and the circle given by $x^2 + y^2 = 10$. Find all solutions to the system of equations. Verify your result both algebraically and graphically.

6. Graph the line given by $x + y = -2$ and the quadratic curve given by $y = x^2 - 4$. Find all solutions to the system of equations. Verify your result both algebraically and graphically.

Lesson Summary

Here are some steps to consider when solving systems of equations that represent a line and a quadratic curve.

1. Solve the linear equation for y in terms of x. This is equivalent to rewriting the equation in slope-intercept form. Note that working with the quadratic equation first would likely be more difficult and might cause the loss of a solution.
2. Replace y in the quadratic equation with the expression involving x from the slope-intercept form of the linear equation. That will yield an equation in one variable.
3. Solve the quadratic equation for x.
4. Substitute x into the linear equation to find the corresponding value of y.
5. Sketch a graph of the system to check your solution.

Problem Set

1. Where do the lines given by $y = x + b$ and $y = 2x + 1$ intersect?

2. Find all solutions to the following system of equations.

$$(x-2)^2 + (y+3)^2 = 4$$
$$x - y = 3$$

Illustrate with a graph.

3. Find all solutions to the following system of equations.

$$x + 2y = 0$$
$$x^2 - 2x + y^2 - 2y - 3 = 0$$

Illustrate with a graph.

4. Find all solutions to the following system of equations.

$$x + y = 4$$
$$(x+3)^2 + (y-2)^2 = 10$$

Illustrate with a graph.

5. Find all solutions to the following system of equations.

$$y = -2x + 3$$
$$y = x^2 - 6x + 3$$

Illustrate with a graph.

6. Find all solutions to the following system of equations.

$$-y^2 + 6y + x - 9 = 0$$
$$6y = x + 27$$

Illustrate with a graph.

7. Find all values of k so that the following system has two solutions.

$$x^2 + y^2 = 25$$
$$y = k$$

Illustrate with a graph.

8. Find all values of k so that the following system has exactly one solution.

$$y = 5 - (x - 3)^2$$
$$y = k$$

Illustrate with a graph.

9. Find all values of k so that the following system has no solutions.

$$x^2 + (y - k)^2 = 36$$
$$y = 5x + k$$

Illustrate with a graph.

This page intentionally left blank

Lesson 32: Graphing Systems of Equations

Classwork

Opening Exercise

Given the line $y = 2x$, is there a point on the line at a distance 3 from $(1, 3)$? Explain how you know.

Draw a graph showing where the point is.

Exercise 1

Solve the system $(x - 1)^2 + (y - 2)^2 = 2^2$ and $y = 2x + 2$.

What are the coordinates of the center of the circle?

What can you say about the distance from the intersection points to the center of the circle?

Using your graphing tool, graph the line and the circle.

Example 1

Rewrite $x^2 + y^2 - 4x + 2y = -1$ by completing the square in both x and y. Describe the circle represented by this equation.

Using your graphing tool, graph the circle.

In contrast, consider the following equation: $x^2 + y^2 - 2x - 8y = -19$

What happens when you use your graphing tool with this equation?

Exercise 2

Consider a circle with radius 5 and another circle with radius 3. Let d represent the distance between the two centers. We want to know how many intersections there are of these two circles for different values of d. Draw figures for each case.

a. What happens if $d = 8$?

b. What happens if $d = 10$?

c. What happens if $d = 1$?

d. What happens if $d = 2$?

e. For which values of d do the circles intersect in exactly one point? Generalize this result to circles of any radius.

f. For which values of d do the circles intersect in two points? Generalize this result to circles of any radius.

g. For which values of d do the circles not intersect? Generalize this result to circles of any radius.

Example 2

Find the distance between the centers of the two circles with equations below, and use that distance to determine in how many points these circles intersect.

$$x^2 + y^2 = 5$$
$$(x-2)^2 + (y-1)^2 = 3$$

Exercise 3

Use the distance formula to show algebraically and graphically that the following two circles do not intersect.

$$(x-1)^2+(y+2)^2=1$$
$$(x+5)^2+(y-4)^2=4$$

Example 3

Point $A(3,2)$ is on a circle whose center is $C(-2,3)$. What is the radius of the circle?

What is the equation of the circle? Graph it.

Use the fact that the tangent at $A(3, 2)$ is perpendicular to the radius at that point to find the equation of the tangent line. Then graph it.

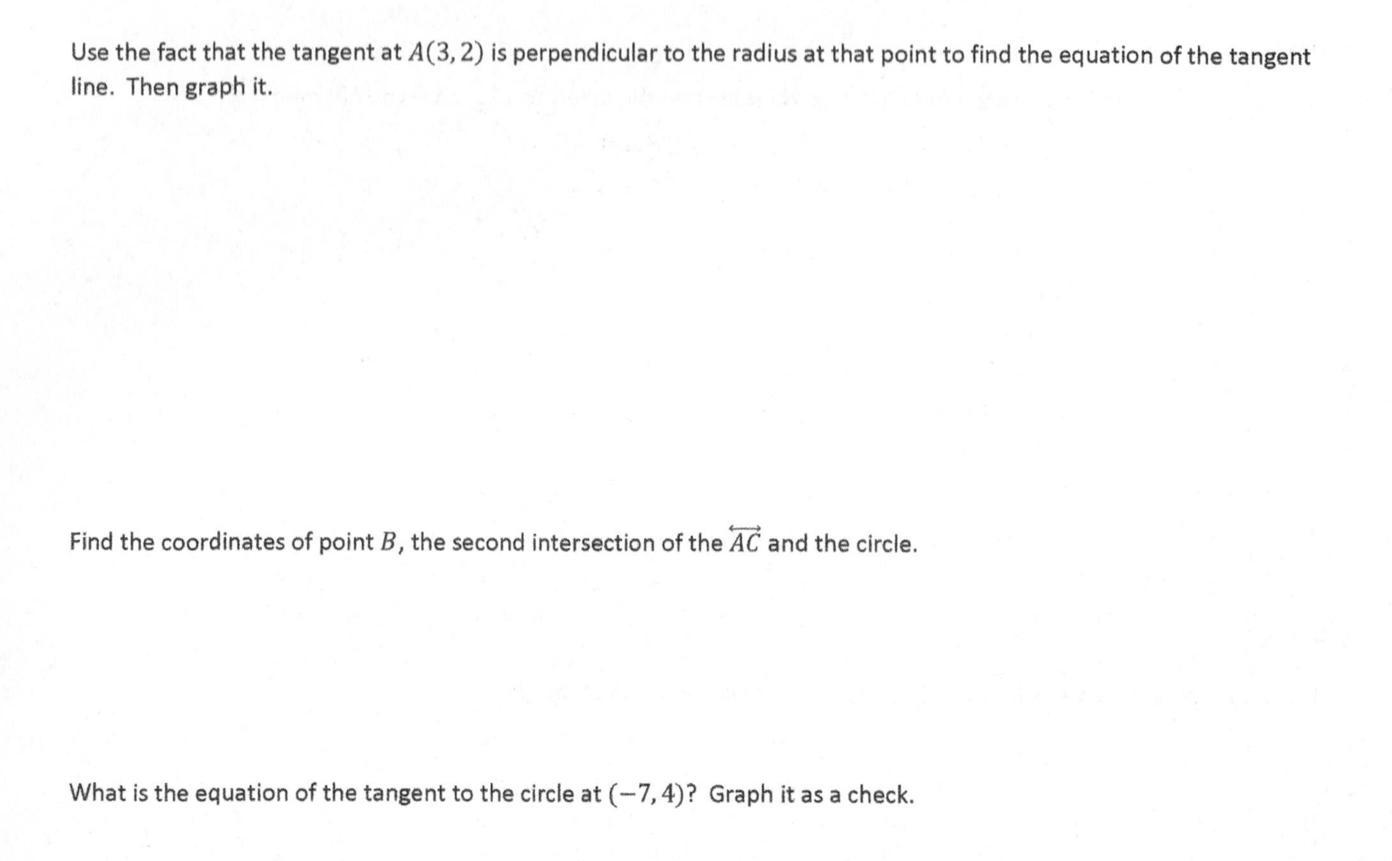

Find the coordinates of point B, the second intersection of the $\overleftrightarrow{AC}$ and the circle.

What is the equation of the tangent to the circle at $(-7, 4)$? Graph it as a check.

The lines $y = 5x + b$ are parallel to the tangent lines to the circle at points A and B. How is the y-intercept b for these lines related to the number of times each line intersects the circle?

Problem Set

1. Use the distance formula to find the distance between the points $(-1, -13)$ and $(3, -9)$.

2. Use the distance formula to find the length of the longer side of the rectangle whose vertices are $(1, 1)$, $(3, 1)$, $(3, 7)$, and $(1, 7)$.

3. Use the distance formula to find the length of the diagonal of the square whose vertices are $(0, 0)$, $(0, 5)$, $(5, 5)$, and $(5, 0)$.

Write an equation for the circles in Exercises 4–6 in the form $(x - h)^2 + (y - k)^2 = r^2$, where the center is (h, k) and the radius is r units. Then write the equation in the standard form $x^2 + ax + y^2 + by + c = 0$, and construct the graph of the equation.

4. A circle with center $(4, -1)$ and radius 6 units.

5. A circle with center $(-3, 5)$ tangent to the x-axis.

6. A circle in the third quadrant, radius 1 unit, tangent to both axes.

7. By finding the radius of each circle and the distance between their centers, show that the circles $x^2 + y^2 = 4$ and $x^2 - 4x + y^2 - 4y + 4 = 0$ intersect. Illustrate graphically.

8. Find the points of intersection of the circles $x^2 + y^2 - 15 = 0$ and $x^2 - 4x + y^2 + 2y - 5 = 0$. Check by graphing the equations.

9. Solve the system $y = x^2 - 2$ and $x^2 + y^2 = 4$. Illustrate graphically.

10. Solve the system $y = 2x - 13$ and $y = x^2 - 6x + 3$. Illustrate graphically.

This page intentionally left blank

Lesson 33: The Definition of a Parabola

Classwork

Opening Exercise

Suppose you are viewing the cross-section of a mirror. Where would the incoming light be reflected in each type of design? Sketch your ideas below.

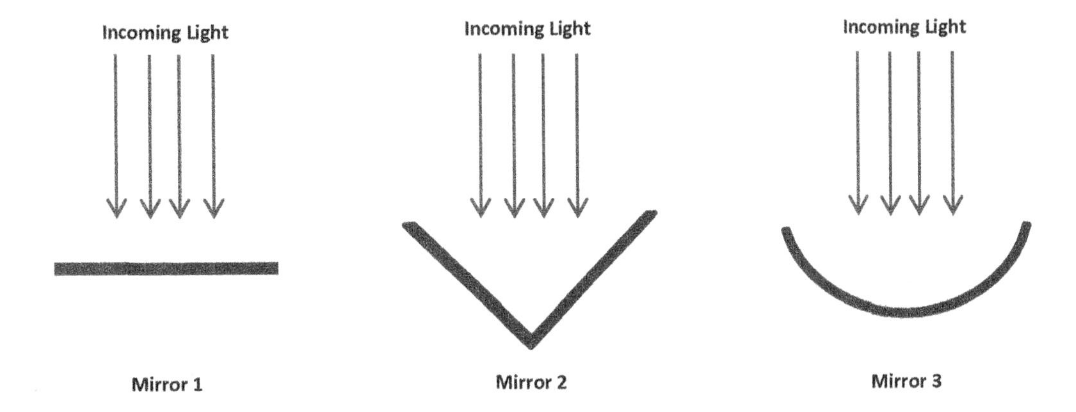

Discussion: Telescope Design

When Newton designed his reflector telescope, he understood two important ideas. Figure 1 shows a diagram of this type of telescope.

- The curved mirror needs to focus all the light to a single point that we will call the focus. An angled flat mirror is placed near this point and reflects the light to the eyepiece of the telescope.
- The reflected light needs to arrive at the focus at the same time. Otherwise, the image is distorted.

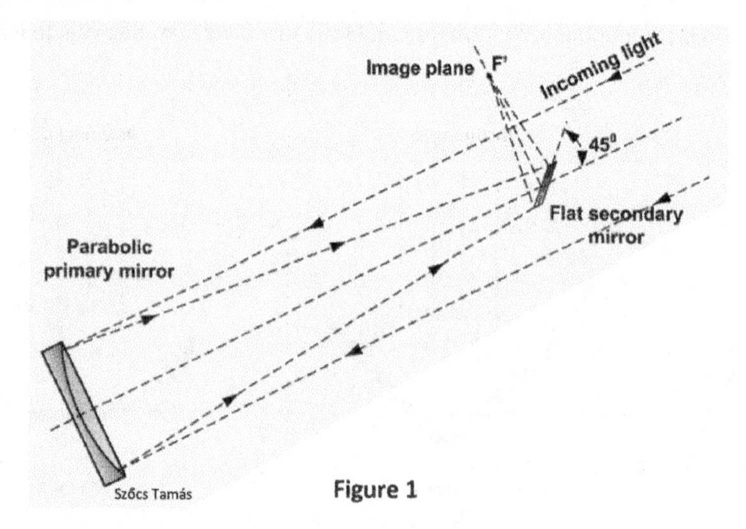

Figure 1

Definition: A *parabola with directrix* L *and focus point* F is the set of all points in the plane that are equidistant from the point F and line L.

Figure 2 to the right illustrates this definition of a parabola. In this diagram, $FP_1 = P_1Q_1, FP_2 = P_2Q_2, FP_3 = P_3Q_3$ showing that for any point P on the parabola, the distance between P and F is equal to the distance between P and the line L.

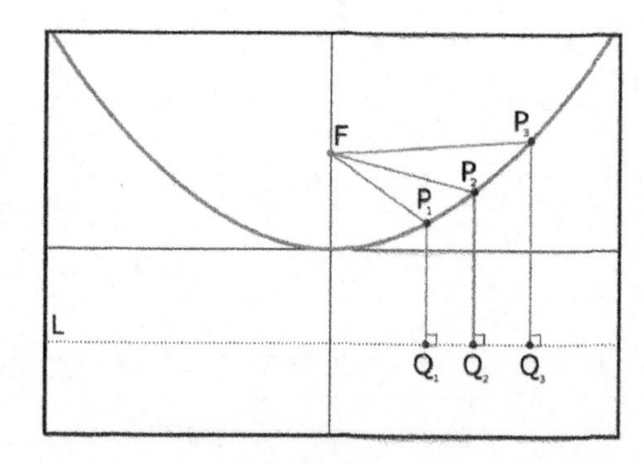

Figure 2

All parabolas have the reflective property illustrated in Figure 3. Rays parallel to the axis reflect off the parabola and through the focus point, F.

Thus, a mirror shaped like a rotated parabola would satisfy Newton's requirements for his telescope design.

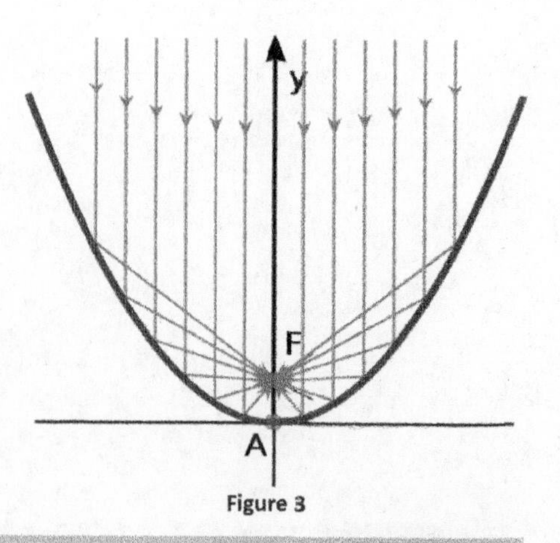

Figure 3

Figure 4 below shows several different line segments representing the reflected light with one endpoint on the curved mirror that is a parabola and the other endpoint at the focus. Anywhere the light hits this type of parabolic surface, it always reflects to the focus, F, at exactly the same time.

Figure 5 shows the same image with a directrix. Imagine for a minute that the mirror was not there. Then, the light would arrive at the directrix all at the same time. Since the distance from each point on the parabolic mirror to the directrix is the same as the distance from the point on the mirror to the focus, and the speed of light is constant, it takes the light the same amount of time to travel to the focus as it would have taken it to travel to the directrix. In the diagram, this means that $AF = AF_A$, $BF = BF_B$, and so on. Thus, the light rays arrive at the focus at the same time, and the image is not distorted.

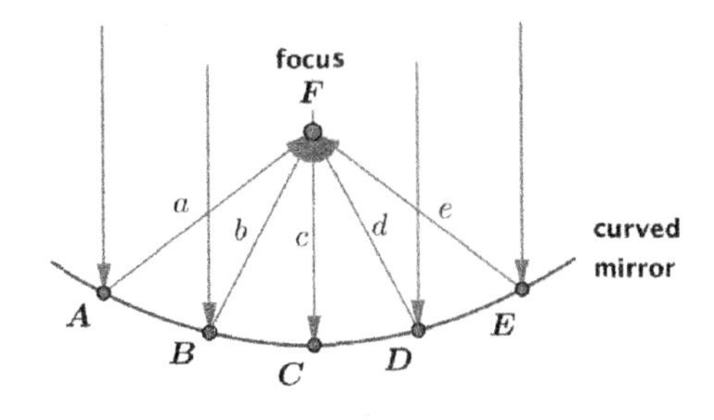

Figure 4

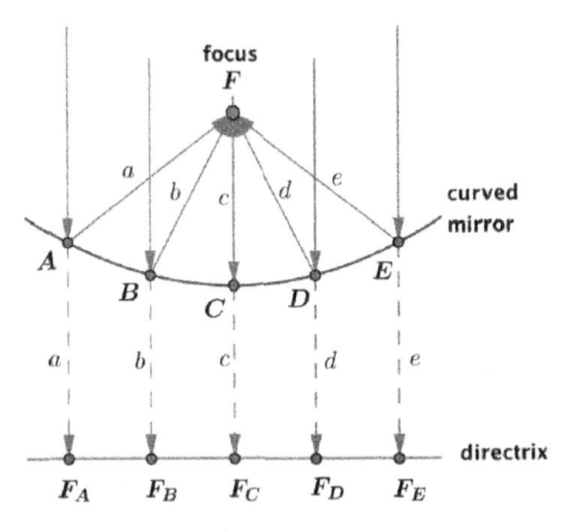

Figure 5

Example: Finding an Analytic Equation for a Parabola

Given a focus and a directrix, create an equation for a parabola.

Focus: $F(0,2)$

Directrix: x-axis

Parabola:
$P = \{(x, y) \mid (x, y)$ is equidistant from F and the x-axis.$\}$

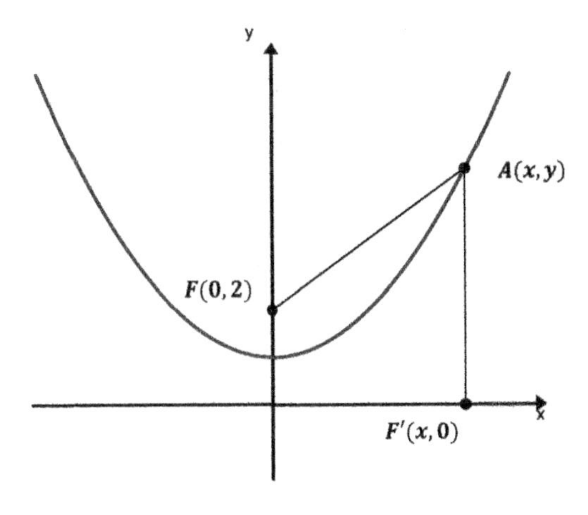

Let A be any point (x, y) on the parabola P. Let F' be a point on the directrix with the same x-coordinate as point A.

What is the length AF'?

Use the distance formula to create an expression that represents the length AF.

Create an equation that relates the two lengths, and solve it for y.

Verify that this equation appears to match the graph shown.

Exercises

1. Demonstrate your understanding of the definition of a parabola by drawing several pairs of congruent segments given the parabola, its focus, and directrix. Measure the segments that you drew to confirm the accuracy of your sketches in either centimeters or inches.

2. Derive the analytic equation of a parabola given the focus of $(0,4)$ and the directrix $y = 2$. Use the diagram to help you work this problem.

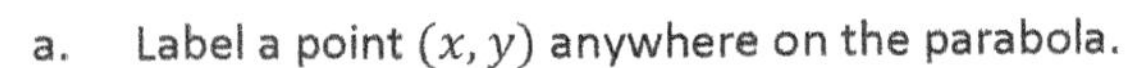

   a. Label a point (x, y) anywhere on the parabola.

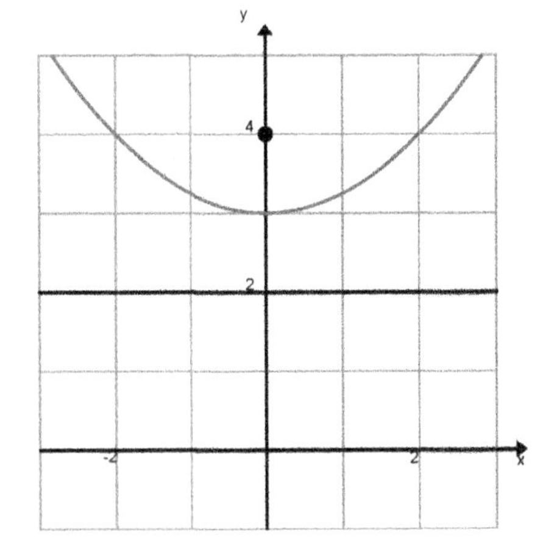

   b. Write an expression for the distance from the point (x, y) to the directrix.

   c. Write an expression for the distance from the point (x, y) to the focus.

   d. Apply the definition of a parabola to create an equation in terms of x and y. Solve this equation for y.

   e. What is the translation that takes the graph of this parabola to the graph of the equation derived in Example 1?

Lesson Summary

PARABOLA: A parabola with directrix line L and focus point F is the set of all points in the plane that are equidistant from the point F and line L

AXIS OF SYMMETRY: The axis of symmetry of a parabola given by a focus point and a directrix is the perpendicular line to the directrix that passes through the focus

VERTEX OF A PARABOLA: The vertex of a parabola is the point where the axis of symmetry intersects the parabola.

In the Cartesian plane, the distance formula can help in deriving an analytic equation for a parabola.

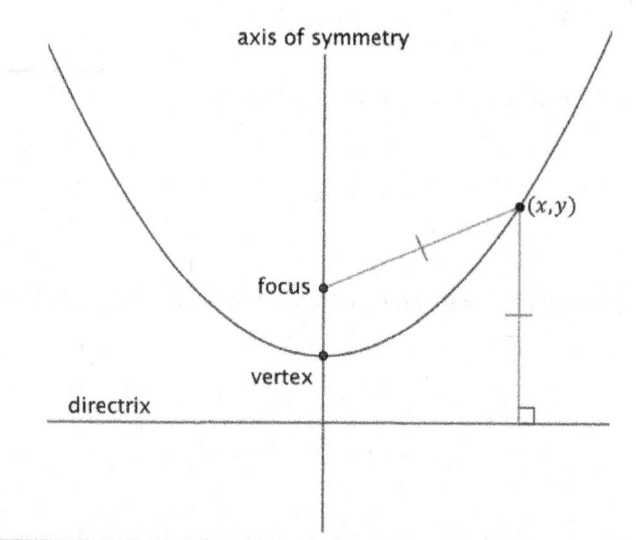

Problem Set

1. Demonstrate your understanding of the definition of a parabola by drawing several pairs of congruent segments given each parabola, its focus, and directrix. Measure the segments that you drew in either inches or centimeters to confirm the accuracy of your sketches.

a.

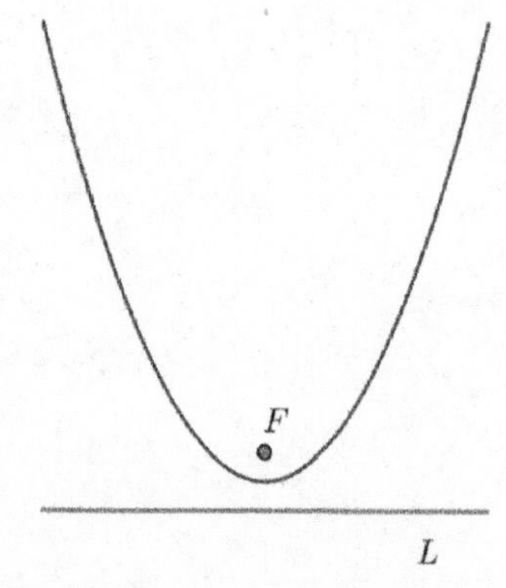

b.

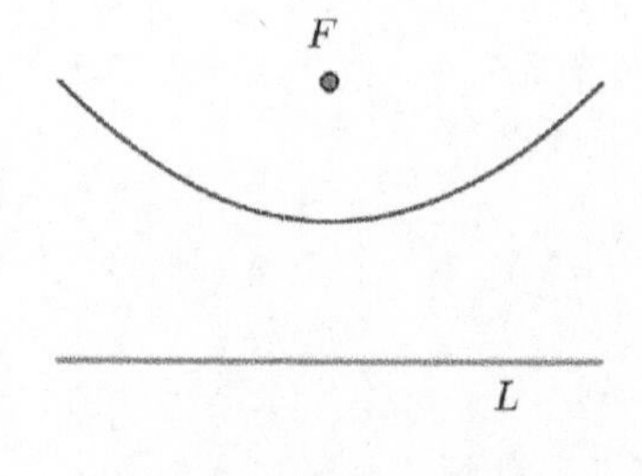

c.

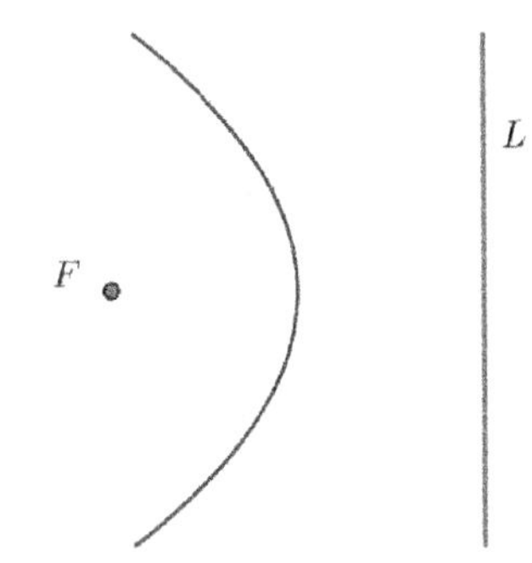

d.

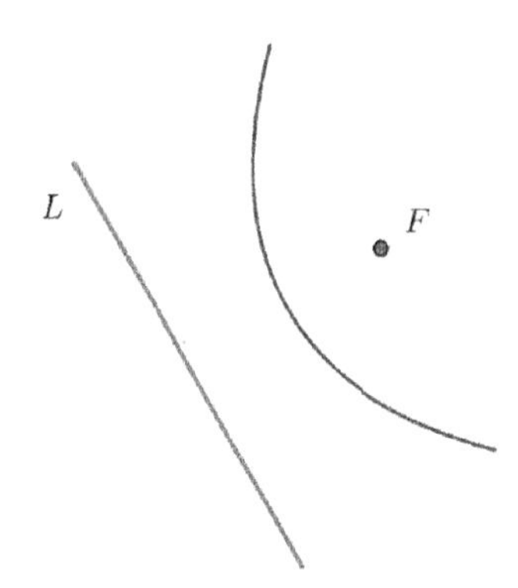

2. Find the distance from the point $(4,2)$ to the point $(0,1)$.

3. Find the distance from the point $(4,2)$ to the line $y = -2$.

4. Find the distance from the point $(-1,3)$ to the point $(3,-4)$.

5. Find the distance from the point $(-1,3)$ to the line $y = 5$.

6. Find the distance from the point $(x, 4)$ to the line $y = -1$.

7. Find the distance from the point $(x, -3)$ to the line $y = 2$.

8. Find the values of x for which the point $(x, 4)$ is equidistant from $(0,1)$, and the line $y = -1$.

9. Find the values of x for which the point $(x, -3)$ is equidistant from $(1, -2)$, and the line $y = 2$.

10. Consider the equation $y = x^2$.
    a. Find the coordinates of the three points on the graph of $y = x^2$ whose x-values are 1, 2, and 3.
    b. Show that each of the three points in part (a) is equidistant from the point $\left(0, \frac{1}{4}\right)$, and the line $y = -\frac{1}{4}$.
    c. Show that if the point with coordinates (x, y) is equidistant from the point $\left(0, \frac{1}{4}\right)$, and the line $y = -\frac{1}{4}$, then $y = x^2$.

11. Consider the equation $y = \frac{1}{2}x^2 - 2x$.

    a. Find the coordinates of the three points on the graph of $y = \frac{1}{2}x^2 - 2x$ whose x-values are -2, 0, and 4.

    b. Show that each of the three points in part (a) is equidistant from the point $\left(2, -\frac{3}{2}\right)$ and the line $y = -\frac{5}{2}$.

    c. Show that if the point with coordinates (x, y) is equidistant from the point $\left(2, -\frac{3}{2}\right)$, and the line $y = -\frac{5}{2}$, then $y = \frac{1}{2}x^2 - 2x$.

12. Derive the analytic equation of a parabola with focus $(1,3)$ and directrix $y = 1$. Use the diagram to help you work this problem.

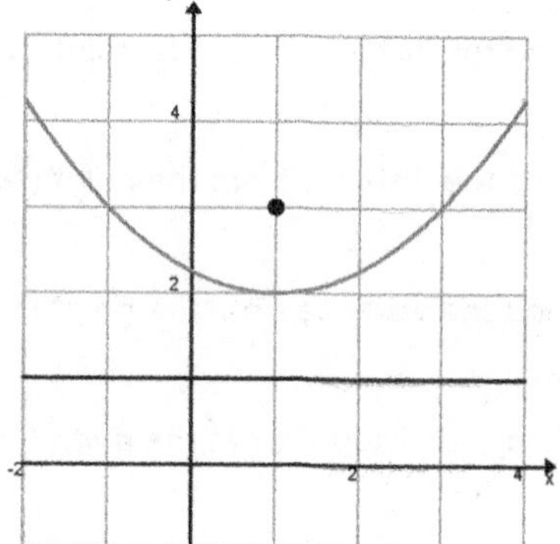

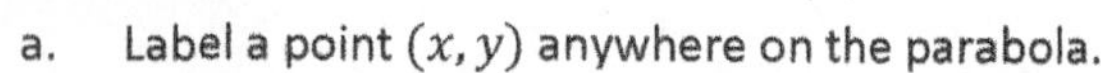

    a. Label a point (x, y) anywhere on the parabola.

    b. Write an expression for the distance from the point (x, y) to the directrix.

    c. Write an expression for the distance from the point (x, y) to the focus $(1,3)$.

    d. Apply the definition of a parabola to create an equation in terms of x and y. Solve this equation for y.

    e. Describe a sequence of transformations that would take this parabola to the parabola with equation $y = \frac{1}{4}x^2 + 1$ derived in Example 1.

13. Consider a parabola with focus $(0, -2)$ and directrix on the x-axis.

    a. Derive the analytic equation for this parabola.

    b. Describe a sequence of transformations that would take the parabola with equation $y = \frac{1}{4}x^2 + 1$ derived in Example 1 to the graph of the parabola in part (a).

14. Derive the analytic equation of a parabola with focus $(0,10)$ and directrix on the x-axis.

Lesson 34: Are All Parabolas Congruent?

Classwork

Opening Exercise

Are all parabolas congruent? Use the following questions to support your answer.

a. Draw the parabola for each focus and directrix given below.

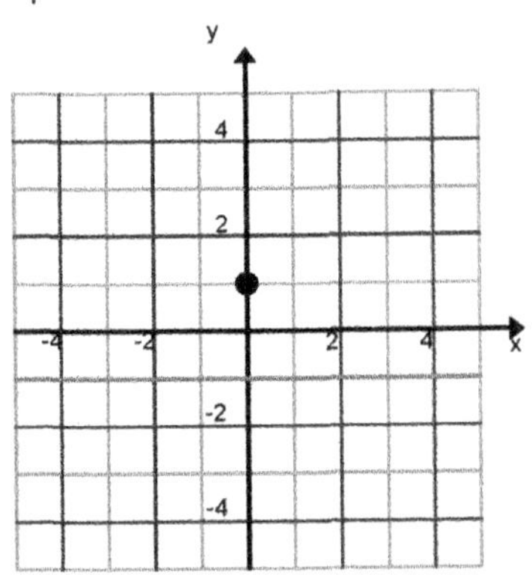

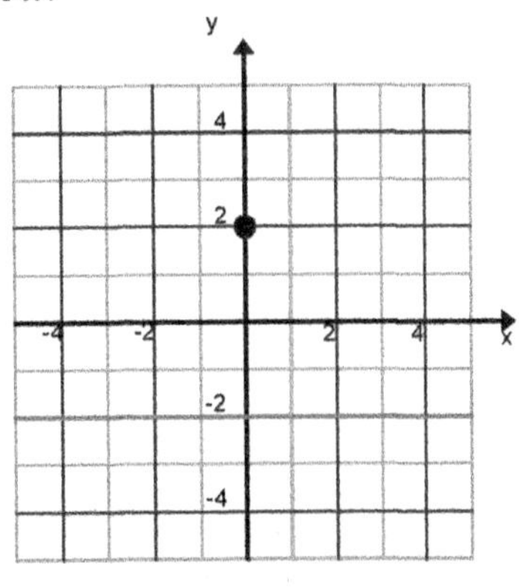

b. What do we mean by congruent parabolas?

c. Are the two parabolas from part (a) congruent? Explain how you know.

d. Are all parabolas congruent?

e. Under what conditions might two parabolas be congruent? Explain your reasoning.

Exercises 1–5

1. Draw the parabola with the given focus and directrix.

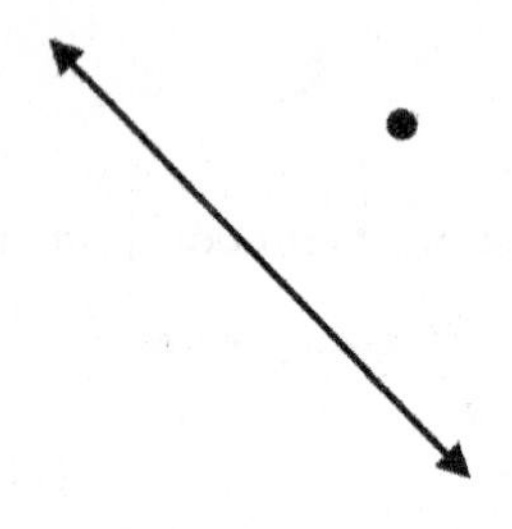

2. Draw the parabola with the given focus and directrix.

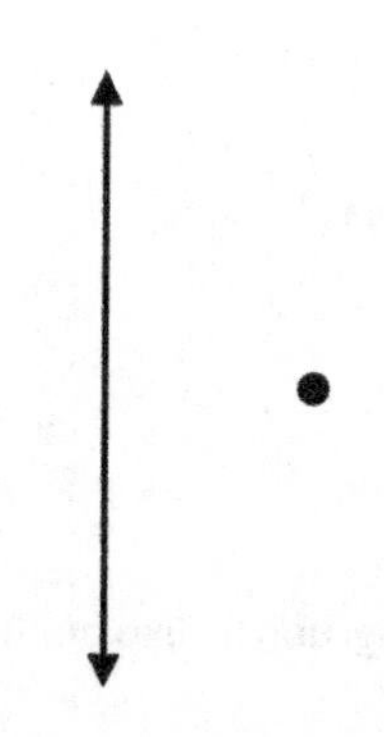

3. Draw the parabola with the given focus and directrix.

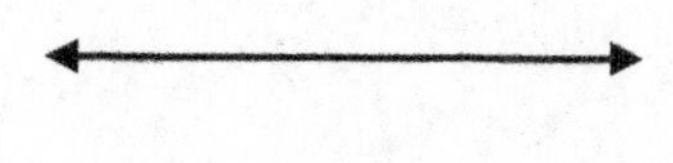

4. What can you conclude about the relationship between the parabolas in Exercises 1–3?

5. Let p be the number of units between the focus and the directrix, as shown. As the value of p increases, what happens to the shape of the resulting parabola?

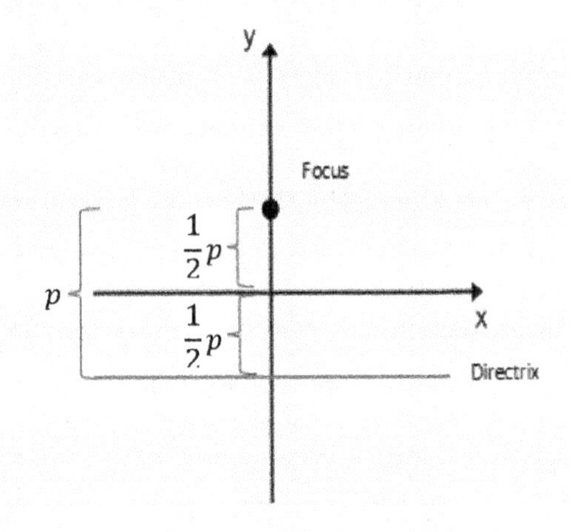

Example 1: Derive an Equation for a Parabola

Consider a parabola P with distance $p > 0$ between the focus with coordinates $\left(0, \frac{1}{2}p\right)$, and directrix $y = -\frac{1}{2}p$.

What is the equation that represents this parabola?

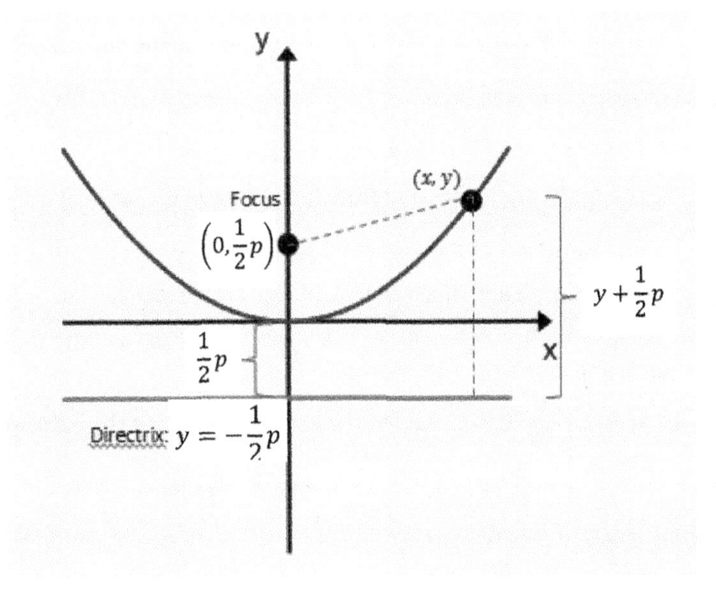

Discussion

We have shown that any parabola with a distance $p > 0$ between the focus $\left(0, \frac{1}{2}p\right)$ and directrix $y = -\frac{1}{2}p$ has a vertex at the origin and is represented by a quadratic equation of the form $y = \frac{1}{2p}x^2$.

Suppose that the vertex of a parabola with a horizontal directrix that opens upward is (h, k), and the distance from the focus to directrix is $p > 0$. Then, the focus has coordinates $\left(h, k + \frac{1}{2}p\right)$, and the directrix has equation $y = k - \frac{1}{2}p$. If we go through the above derivation with focus $\left(h, k + \frac{1}{2}p\right)$ and directrix $y = k - \frac{1}{2}p$, we should not be surprised to get a quadratic equation. In fact, if we complete the square on that equation, we can write it in the form $y = \frac{1}{2p}(x - h)^2 + k$.

In Algebra I, Module 4, Topic B, we saw that any quadratic function can be put into vertex form: $f(x) = a(x - h)^2 + k$. Now we see that any parabola that opens upward can be described by a quadratic function in vertex form, where $a = \frac{1}{2p}$.

If the parabola opens downward, then the equation is $y = -\frac{1}{2p}(x - h)^2 + k$, and the graph of any quadratic equation of this form is a parabola with vertex at (h, k), distance p between focus and directrix, and opening downward. Likewise, we can derive analogous equations for parabolas that open to the left and right. This discussion is summarized in the box below.

Vertex Form of a Parabola

Given a parabola P with vertex (h, k), horizontal directrix, and distance $p > 0$ between focus and directrix, the analytic equation that describes the parabola P is

- $y = \frac{1}{2p}(x - h)^2 + k$ if the parabola opens upward, and
- $y = -\frac{1}{2p}(x - h)^2 + k$ if the parabola opens downward.

Conversely, if $p > 0$, then

- The graph of the quadratic equation $y = \frac{1}{2p}(x - h)^2 + k$ is a parabola that opens upward with vertex at (h, k) and distance p from focus to directrix, and
- The graph of the quadratic equation $y = -\frac{1}{2p}(x - h)^2 + k$ is a parabola that opens downward with vertex at (h, k) and distance p from focus to directrix.

Given a parabola P with vertex (h, k), vertical directrix, and distance $p > 0$ between focus and directrix, the analytic equation that describes the parabola P is

- $x = \frac{1}{2p}(y - k)^2 + h$ if the parabola opens to the right, and
- $x = -\frac{1}{2p}(y - k)^2 + h$ if the parabola opens to the left.

Conversely, if $p > 0$, then

- The graph of the quadratic equation $x = \frac{1}{2p}(y - k)^2 + h$ is a parabola that opens to the right with vertex at (h, k) and distance p from focus to directrix, and
- The graph of the quadratic equation $x = -\frac{1}{2p}(y - k)^2 + h$ is a parabola that opens to the left with vertex at (h, k) and distance p from focus to directrix.

Example 2

THEOREM: Given a parabola P given by a directrix L and a focus F in the Cartesian plane, then P is congruent to the graph of $y = \frac{1}{2p}x^2$, where p is the distance from F to L.

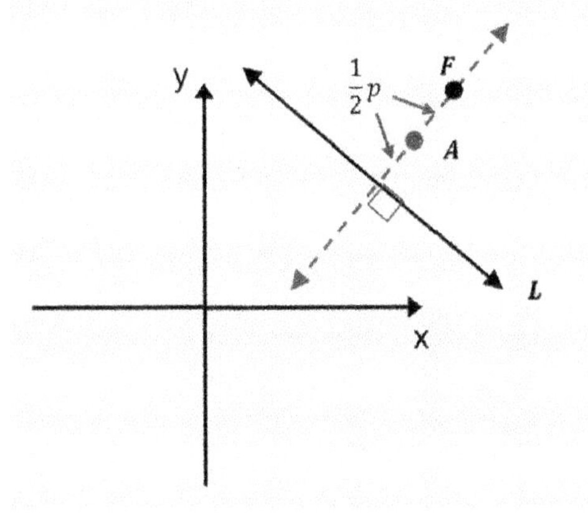

PROOF

Exercises 6–9: Reflecting on the Theorem

6. Restate the results of the theorem from Example 2 in your own words.

7. Create the equation for a parabola that is congruent to $y = 2x^2$. Explain how you determined your answer.

8. Create an equation for a parabola that IS NOT congruent to $y = 2x^2$. Explain how you determined your answer.

9. Write the equation for two different parabolas that are congruent to the parabola with focus point $(0,3)$ and directrix line $y = -3$.

Problem Set

1. Show that if the point with coordinates (x, y) is equidistant from $(4,3)$, and the line $y = 5$, then $y = -\frac{1}{4}x^2 + 2x$.

2. Show that if the point with coordinates (x, y) is equidistant from the point $(2,0)$ and the line $y = -4$, then $y = \frac{1}{8}(x-2)^2 - 2$.

3. Find the equation of the set of points which are equidistant from $(0,2)$ and the x-axis. Sketch this set of points.

4. Find the equation of the set of points which are equidistant from the origin and the line $y = 6$. Sketch this set of points.

5. Find the equation of the set of points which are equidistant from $(4, -2)$ and the line $y = 4$. Sketch this set of points.

6. Find the equation of the set of points which are equidistant from $(4,0)$ and the y-axis. Sketch this set of points.

7. Find the equation of the set of points which are equidistant from the origin and the line $x = -2$. Sketch this set of points.

8. Use the definition of a parabola to sketch the parabola defined by the given focus and directrix.
   a. Focus: $(0,5)$ Directrix: $y = -1$
   b. Focus: $(-2,0)$ Directrix: y-axis
   c. Focus: $(4, -4)$ Directrix: x-axis
   d. Focus: $(2,4)$ Directrix: $y = -2$

9. Find an analytic equation for each parabola described in Problem 8.

10. Are any of the parabolas described in Problem 9 congruent? Explain your reasoning.

11. Sketch each parabola, labeling its focus and directrix.
    a. $y = \frac{1}{2}x^2 + 2$
    b. $y = -\frac{1}{4}x^2 + 1$
    c. $x = \frac{1}{8}y^2$
    d. $x = \frac{1}{2}y^2 + 2$
    e. $y = \frac{1}{10}(x-1)^2 - 2$

12. Determine which parabolas are congruent to the parabola with equation $y = -\frac{1}{4}x^2$.

a.

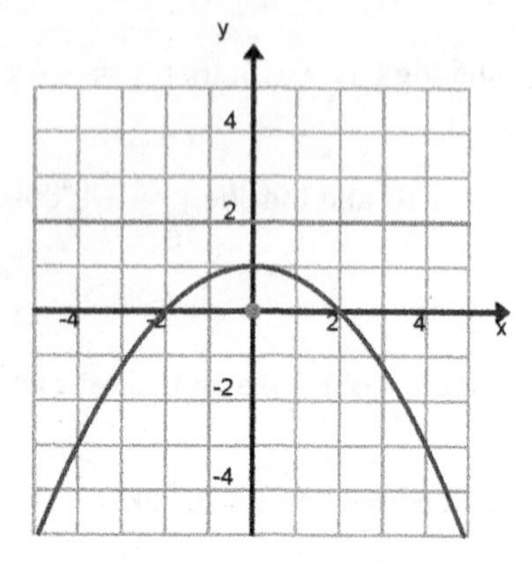

c.

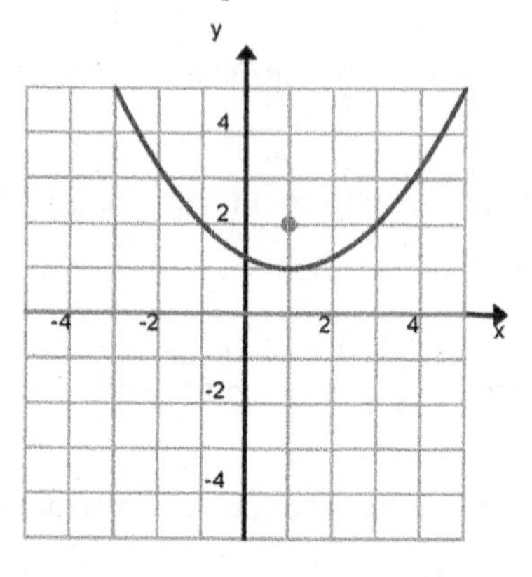

b.

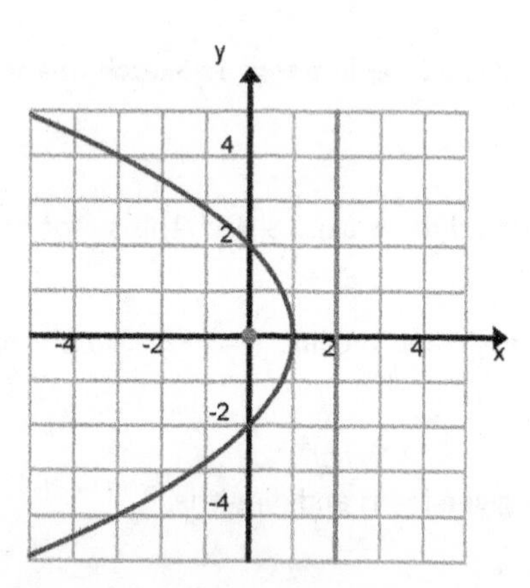

d.

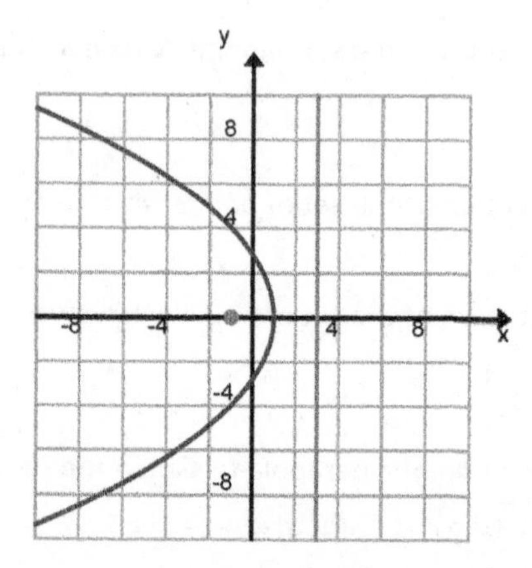

13. Determine which equations represent the graph of a parabola that is congruent to the parabola shown to right.

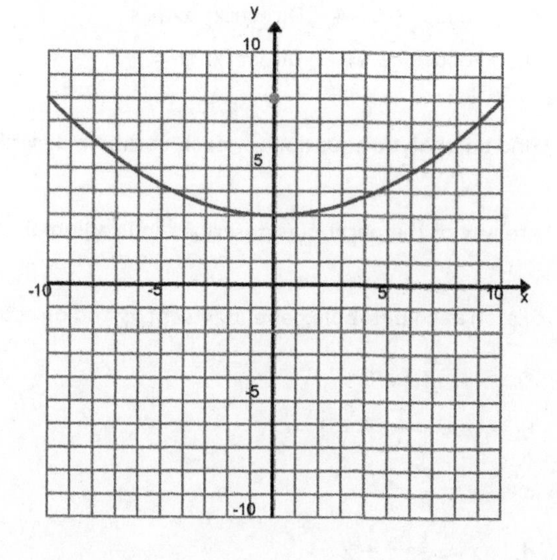

a. $y = \frac{1}{20}x^2$

b. $y = \frac{1}{10}x^2 + 3$

c. $y = -\frac{1}{20}x^2 + 8$

d. $y = \frac{1}{5}x^2 + 5$

e. $x = \frac{1}{10}y^2$

f. $x = \frac{1}{5}(y - 3)^2$

g. $x = \frac{1}{20}y^2 + 1$

14. Jemma thinks that the parabola with equation $y = \frac{1}{3}x^2$ is NOT congruent to the parabola with equation $y = -\frac{1}{3}x^2 + 1$. Do you agree or disagree? Create a convincing argument to support your reasoning.

15. Let P be the parabola with focus $(2,6)$ and directrix $y = -2$.
    a. Write an equation whose graph is a parabola congruent to P with focus $(0,4)$.
    b. Write an equation whose graph is a parabola congruent to P with focus $(0,0)$.
    c. Write an equation whose graph is a parabola congruent to P with the same directrix but different focus.
    d. Write an equation whose graph is a parabola congruent to P with the same focus but with a vertical directrix.

16. Let P be the parabola with focus $(0,4)$ and directrix $y = x$.
    a. Sketch this parabola.
    b. By how many degrees would you have to rotate P about the focus to make the directrix line horizontal?
    c. Write an equation in the form $y = \frac{1}{2a}x^2$ whose graph is a parabola that is congruent to P.
    d. Write an equation whose graph is a parabola with a vertical directrix that is congruent to P.
    e. Write an equation whose graph is P', the parabola congruent to P that results after P is rotated clockwise $45°$ about the focus.
    f. Write an equation whose graph is P'', the parabola congruent to P that results after the directrix of P is rotated $45°$ about the origin.

Extension:

17. Consider the function $f(x) = \frac{2x^2-8x+9}{-x^2+4x-5}$, where x is a real number.
    a. Use polynomial division to rewrite f in the form $f(x) = q + \frac{r}{-x^2+4x-5}$ for some real numbers q and r.
    b. Find the x-value where the maximum occurs for the function f without using graphing technology. Explain how you know.

This page intentionally left blank

Lesson 35: Are All Parabolas Similar?

Classwork

Exercises 1–8

1. Write equations for two parabolas that are congruent to the parabola given by $y = x^2$, and explain how you determined your equations.

2. Sketch the graph of $y = x^2$ and the two parabolas you created on the same coordinate axes.

3. Write the equation of two parabolas that are NOT congruent to $y = x^2$. Explain how you determined your equations.

4. Sketch the graph of $y = x^2$ and the two non-congruent parabolas you created on the same coordinate axes.

5. What does it mean for two triangles to be similar? How do we use geometric transformations to determine if two triangles are similar?

6. What would it mean for two parabolas to be similar? How could we use geometric transformation to determine if two parabolas are similar?

7. Use your work in Exercises 1–6 to make a conjecture: Are all parabolas similar? Explain your reasoning.

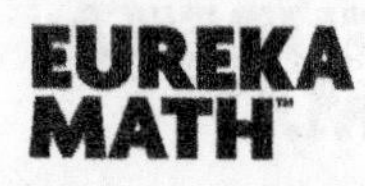

8. The parabola at right is the graph of which equation?

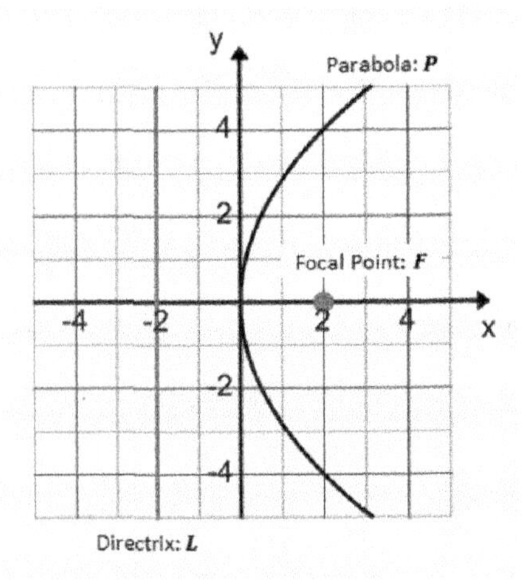

   a. Label a point (x, y) on the graph of P.

   b. What does the definition of a parabola tell us about the distance between the point (x, y) and the directrix L, and the distance between the point (x, y) and the focus F?

   c. Create an equation that relates these two distances.

   d. Solve this equation for x.

   e. Find two points on the parabola P, and show that they satisfy the equation found in part (d).

Discussion

Do you think that all parabolas are similar? Explain why you think so.

What could we do to show that two parabolas are similar? How might you show this?

Exercises 9–12

Use the graphs below to answer Exercises 9 and 10.

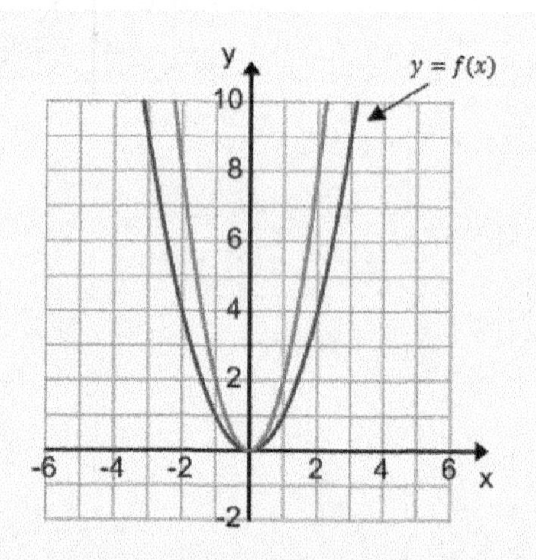

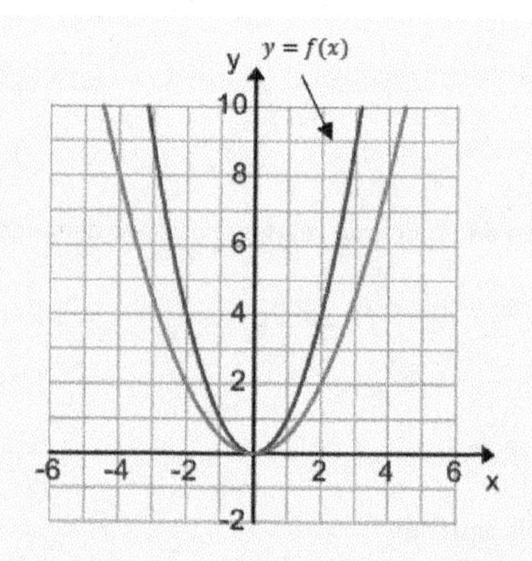

9. Suppose the unnamed red graph on the left coordinate plane is the graph of a function g. Describe g as a vertical scaling of the graph of $y = f(x)$; that is, find a value of k so that $g(x) = kf(x)$. What is the value of k? Explain how you determined your answer.

10. Suppose the unnamed red graph on the right coordinate plane is the graph of a function h. Describe h as a vertical scaling of the graph of $y = f(x)$; that is, find a value of k so that $h(x) = kf(x)$. Explain how you determined your answer.

Use the graphs below to answer Exercises 11–12.

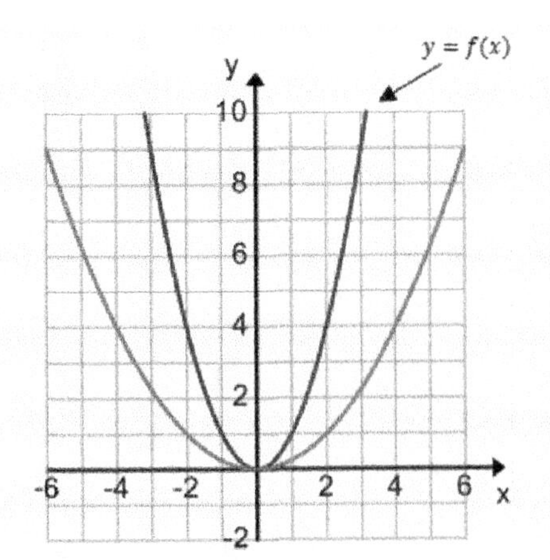

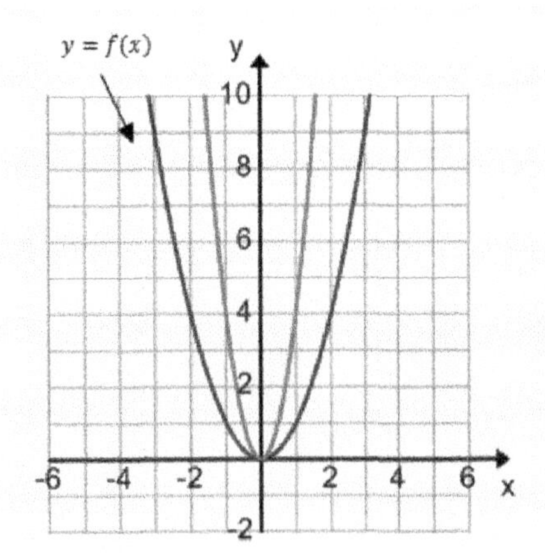

11. Suppose the unnamed function graphed in red on the left coordinate plane is g. Describe g as a horizontal scaling of the graph of $y = f(x)$. What is the value of the scale factor k? Explain how you determined your answer.

12. Suppose the unnamed function graphed in red on the right coordinate plane is h. Describe h as a horizontal scaling of the graph of $y = f(x)$. What is the value of the scale factor k? Explain how you determined your answer.

Definition: A *dilation at the origin* D_k is a horizontal scaling by $k > 0$ followed by a vertical scaling by the same factor k. In other words, this dilation of the graph of $y = f(x)$ is the graph of the equation $y = kf\left(\frac{1}{k}x\right)$.

Example: Dilation at the Origin

Let $f(x) = x^2$ and let $k = 2$. Write a formula for the function g that results from dilating f at the origin by a factor of $\frac{1}{2}$.

What would the results be for $k = 3, 4,$ or 5? What about $k = \frac{1}{2}$?

Lesson Summary

- We started with a geometric figure of a parabola defined by geometric requirements and recognized that it involved the graph of an equation we studied in algebra.
- We used algebra to prove that all parabolas with the same distance between the focus and directrix are congruent to each other, and in particular, they are congruent to a parabola with vertex at the origin, axis of symmetry along the y-axis, and equation of the form $y = \frac{1}{2p}x^2$.
- Noting that the equation for a parabola with axis of symmetry along the y-axis is of the form $y = f(x)$ for a quadratic function f, we proved that all parabolas are similar using transformations of functions.

Problem Set

1. Let $(x) = \sqrt{4 - x^2}$. The graph of f is shown below. On the same axes, graph the function g, where $g(x) = f\left(\frac{1}{2}x\right)$. Then, graph the function h, where $h(x) = 2g(x)$.

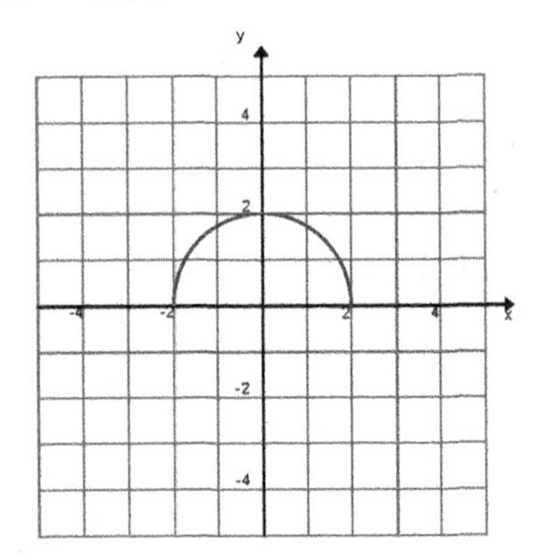

2. Let $f(x) = -|x| + 1$. The graph of f is shown below. On the same axes, graph the function g, where $g(x) = f\left(\frac{1}{3}x\right)$. Then, graph the function h, where $h(x) = 3g(x)$.

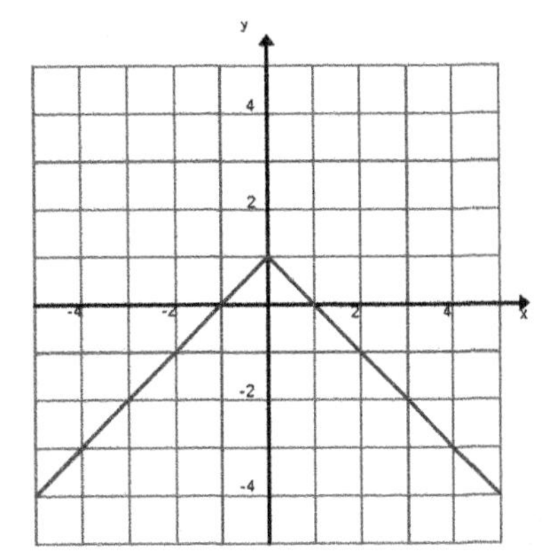

3. Based on your work in Problems 1 and 2, describe the resulting function when the original function is transformed with a horizontal and then a vertical scaling by the same factor, k.

4. Let $f(x) = x^2$.
   a. What are the focus and directrix of the parabola that is the graph of the function $f(x) = x^2$?
   b. Describe the sequence of transformations that would take the graph of f to each parabola described below.
      i. Focus: $\left(0, -\frac{1}{4}\right)$, directrix: $y = \frac{1}{4}$
      ii. Focus: $\left(\frac{1}{4}, 0\right)$, directrix: $x = -\frac{1}{4}$
      iii. Focus: $(0, 0)$, directrix: $y = -\frac{1}{2}$
      iv. Focus: $\left(0, \frac{1}{4}\right)$, directrix: $y = -\frac{3}{4}$
      v. Focus: $(0, 3)$, directrix: $y = -1$
   c. Which parabolas are similar to the parabola that is the graph of f? Which are congruent to the parabola that is the graph of f?

5. Derive the analytic equation for each parabola described in Problem 4(b) by applying your knowledge of transformations.

6. Are all parabolas the graph of a function of x in the xy-plane? If so, explain why, and if not, provide an example (by giving a directrix and focus) of a parabola that is not.

7. Are the following parabolas congruent? Explain your reasoning.

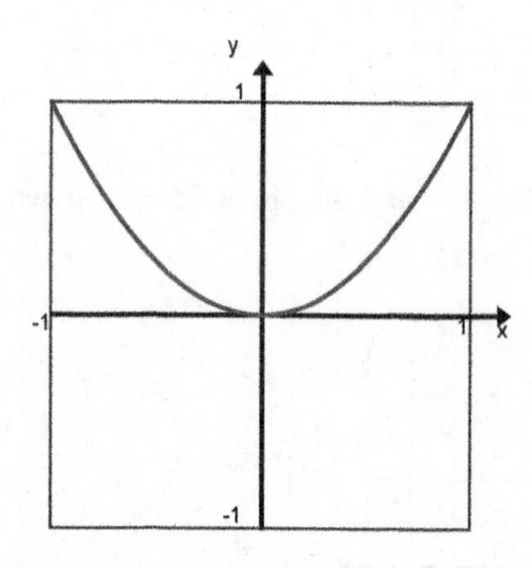

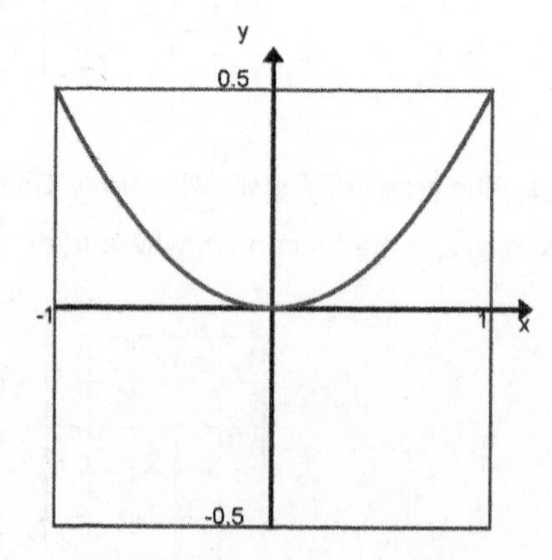

8. Are the following parabolas congruent? Explain your reasoning.

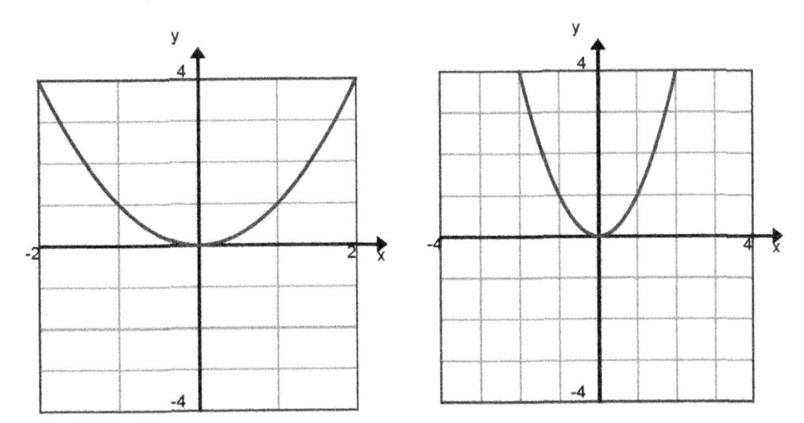

9. Write the equation of a parabola congruent to $y = 2x^2$ that contains the point $(1, -2)$. Describe the transformations that would take this parabola to your new parabola.

10. Write the equation of a parabola similar to $y = 2x^2$ that does NOT contain the point $(0,0)$ but does contain the point $(1,1)$.

This page intentionally left blank

Lesson 36: Overcoming a Third Obstacle to Factoring—What If There Are No Real Number Solutions?

Classwork

Opening Exercise

Find all solutions to each of the systems of equations below using any method.

$2x - 4y = -1$ $3x - 6y = 4$	$y = x^2 - 2$ $y = 2x - 5$	$x^2 + y^2 = 1$ $x^2 + y^2 = 4$

Exercises 1–4

1. Are there any real number solutions to the system $y = 4$ and $x^2 + y^2 = 2$? Support your findings both analytically and graphically.

2. Does the line $y = x$ intersect the parabola $y = -x^2$? If so, how many times and where? Draw graphs on the same set of axes.

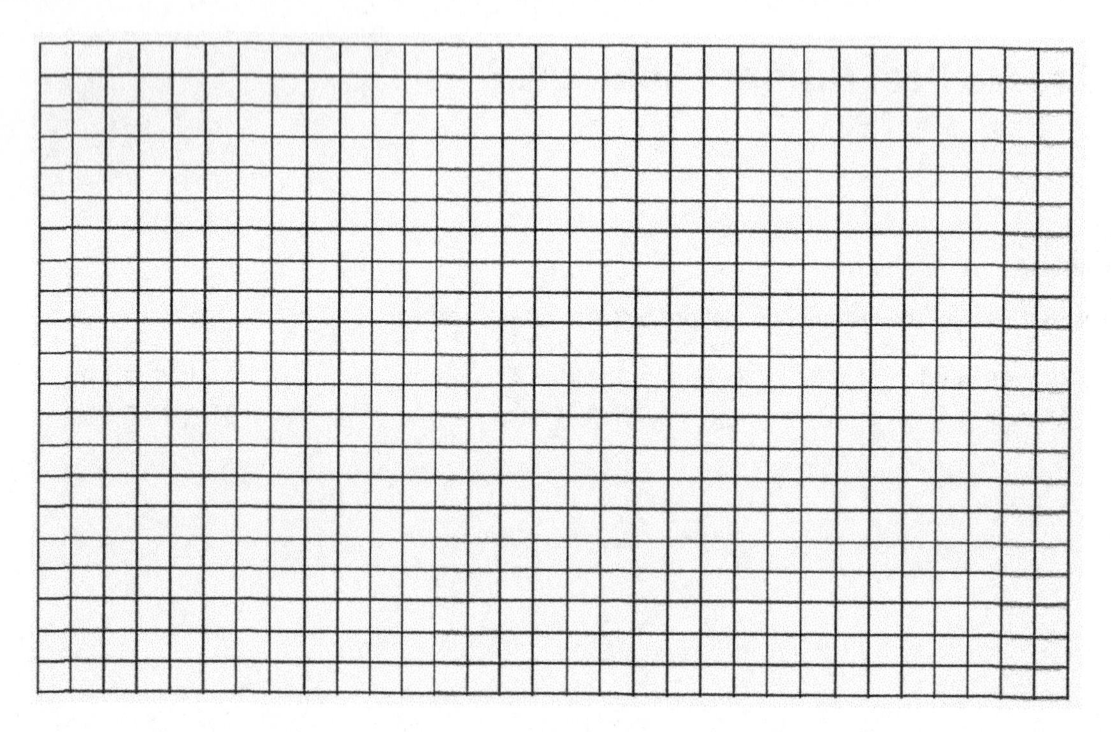

3. Does the line $y = -x$ intersect the circle $x^2 + y^2 = 1$? If so, how many times and where? Draw graphs on the same set of axes.

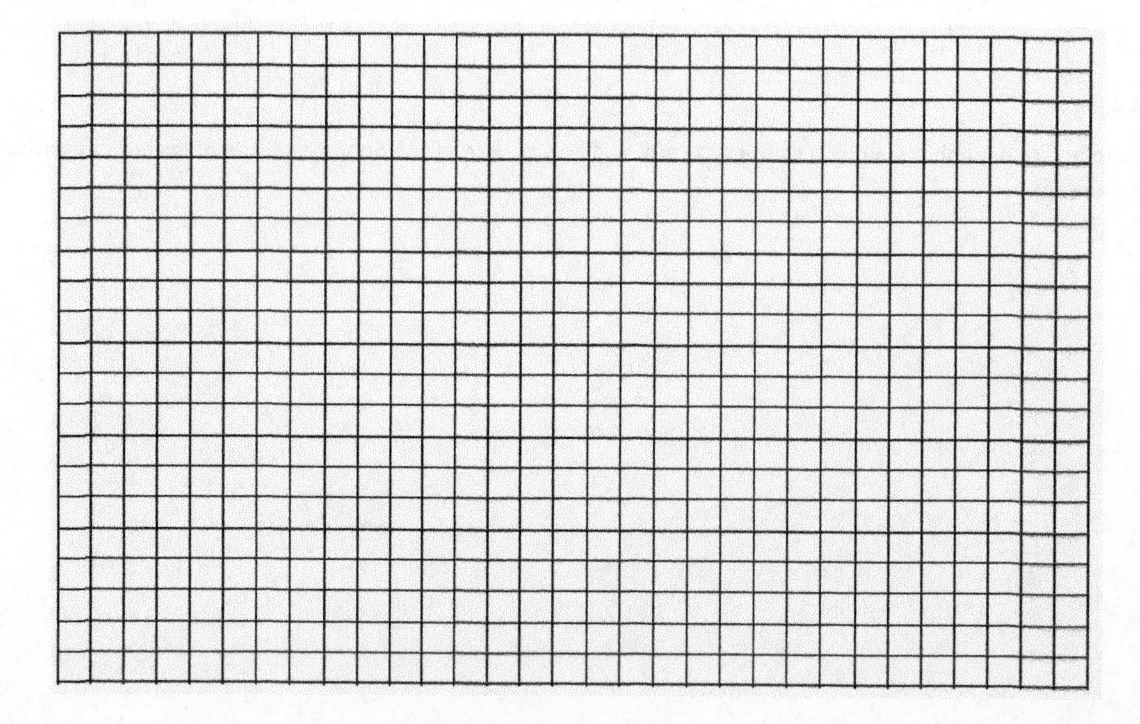

4. Does the line $y = 5$ intersect the parabola $y = 4 - x^2$? Why or why not? Draw the graphs on the same set of axes.

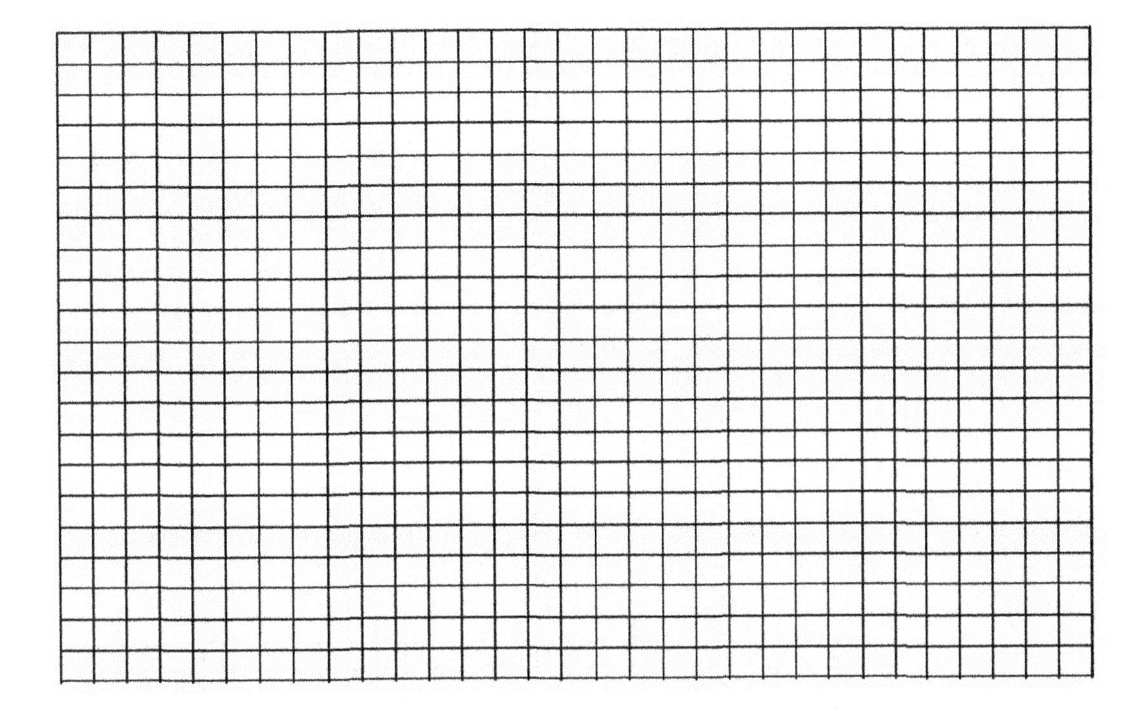

Lesson Summary

An equation or a system of equations may have one or more solutions in the real numbers, or it may have no real number solution.

Two graphs that do not intersect in the coordinate plane correspond to a system of two equations without a real solution. If a system of two equations does not have a real solution, the graphs of the two equations do not intersect in the coordinate plane.

A quadratic equation in the form $ax^2 + bx + c = 0$, where a, b, and c are real numbers and $a \neq 0$, that has no real solution indicates that the graph of $y = ax^2 + bx + c$ does not intersect the x-axis.

Problem Set

1. For each part, solve the system of linear equations, or show that no real solution exists. Graphically support your answer.

   a. $4x + 2y = 9$
      $x + y = 3$

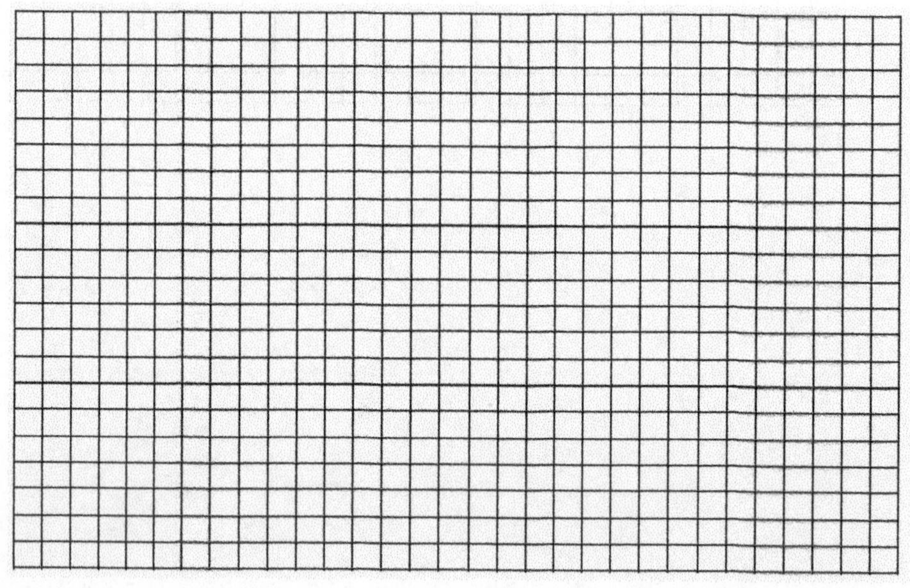

b. $2x - 8y = 9$
$3x - 12y = 0$

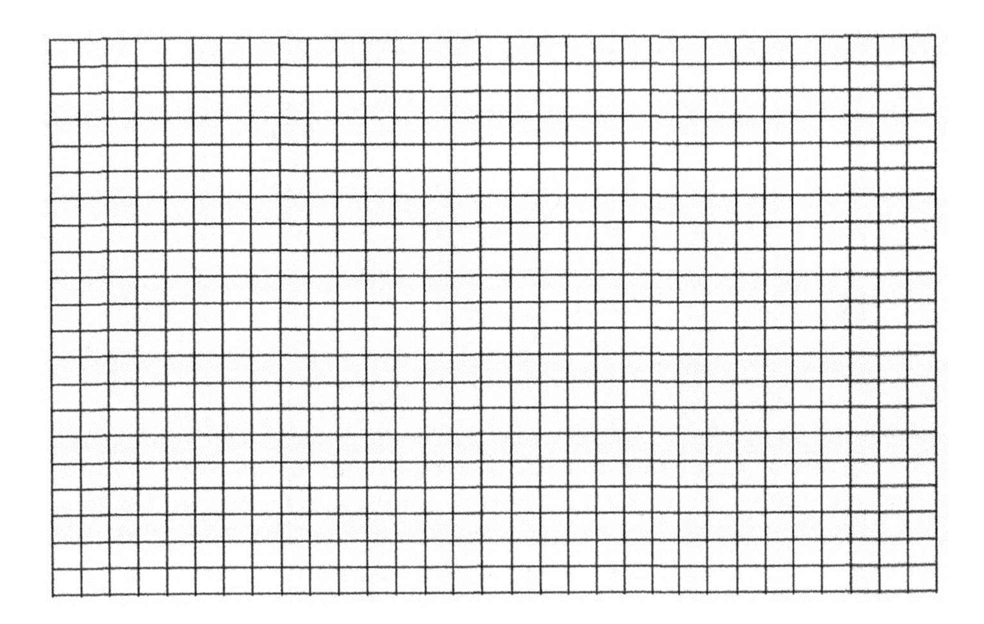

2. Solve the following system of equations, or show that no real solution exists. Graphically confirm your answer.

$$3x^2 + 3y^2 = 6$$
$$x - y = 3$$

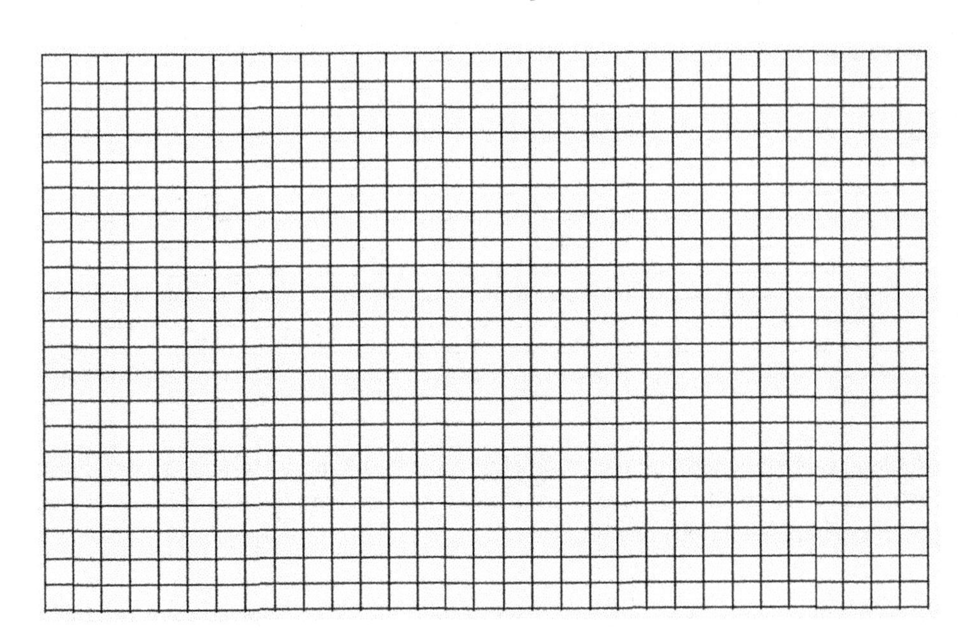

3. Find the value of k so that the graph of the following system of equations has no solution.

$$3x - 2y - 12 = 0$$
$$kx + 6y - 10 = 0$$

4. Offer a geometric explanation to why the equation $x^2 - 6x + 10 = 0$ has no real solutions.

5. Without his pencil or calculator, Joey knows that $2x^3 + 3x^2 - 1 = 0$ has at least one real solution. How does he know?

6. The graph of the quadratic equation $y = x^2 + 1$ has no x-intercepts. However, Gia claims that when the graph of $y = x^2 + 1$ is translated by a distance of 1 in a certain direction, the new (translated) graph would have exactly one x-intercept. Further, if $y = x^2 + 1$ is translated by a distance greater than 1 in the same direction, the new (translated) graph would have exactly two x-intercepts. Support or refute Gia's claim. If you agree with her, in which direction did she translate the original graph? Draw graphs to illustrate.

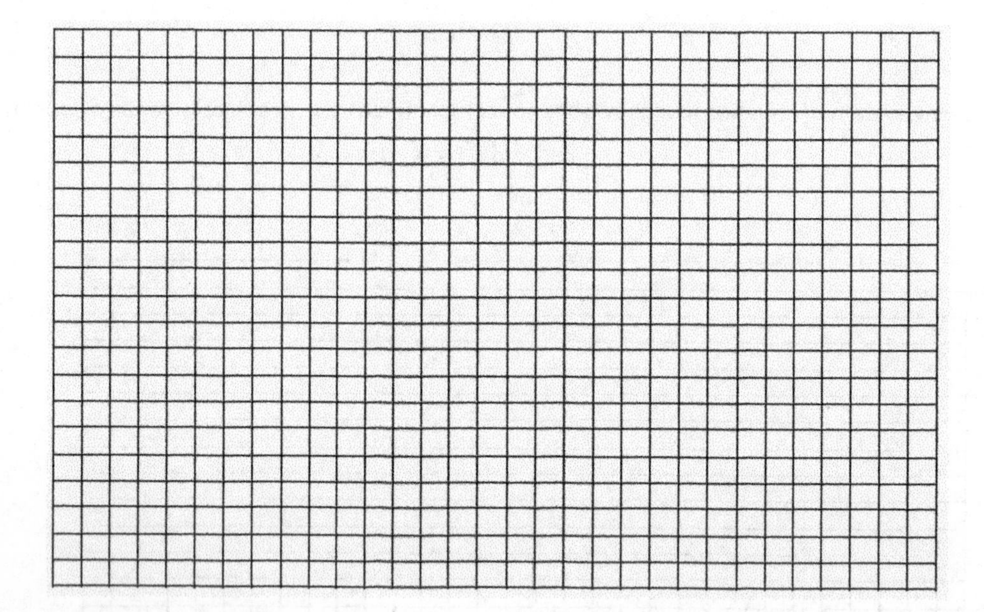

7. In the previous problem, we mentioned that the graph of $y = x^2 + 1$ has no x-intercepts. Suppose that $y = x^2 + 1$ is one of two equations in a system of equations and that the other equation is linear. Give an example of a linear equation such that this system has exactly one solution.

8. In prior problems, we mentioned that the graph of $y = x^2 + 1$ has no x-intercepts. Does the graph of $y = x^2 + 1$ intersect the graph of $y = x^3 + 1$?

Lesson 37: A Surprising Boost from Geometry

Classwork

Opening Exercise

Solve each equation for x.

a. $x - 1 = 0$

b. $x + 1 = 0$

c. $x^2 - 1 = 0$

d. $x^2 + 1 = 0$

Example 1: Addition with Complex Numbers

Compute $(3 + 4i) + (7 - 20i)$.

Example 2: Subtraction with Complex Numbers

Compute $(3 + 4i) - (7 - 20i)$.

Example 3: Multiplication with Complex Numbers

Compute $(1 + 2i)(1 - 2i)$.

Example 4: Multiplication with Complex Numbers

Verify that $-1 + 2i$ and $-1 - 2i$ are solutions to $x^2 + 2x + 5 = 0$.

Lesson Summary

Multiplication by i rotates every complex number in the complex plane by $90°$ about the origin.

Every complex number is in the form $a + bi$, where a is the real part and b is the imaginary part of the number. Real numbers are also complex numbers; the real number a can be written as the complex number $a + 0i$. Numbers of the form bi, for real numbers b, are called imaginary numbers.

Adding two complex numbers is analogous to combining like terms in a polynomial expression.

Multiplying two complex numbers is like multiplying two binomials, except one can use $i^2 = -1$ to further write the expression in simpler form.

Complex numbers satisfy the associative, commutative, and distributive properties.

Complex numbers allow us to find solutions to polynomial equations that have no real number solutions.

Problem Set

1. Locate the point on the complex plane corresponding to the complex number given in parts (a)–(h). On one set of axes, label each point by its identifying letter. For example, the point corresponding to $5 + 2i$ should be labeled a.

   a. $5 + 2i$

   b. $3 - 2i$

   c. $-2 - 4i$

   d. $-i$

   e. $\frac{1}{2} + i$

   f. $\sqrt{2} - 3i$

   g. 0

   h. $-\frac{3}{2} + \frac{\sqrt{3}}{2}i$

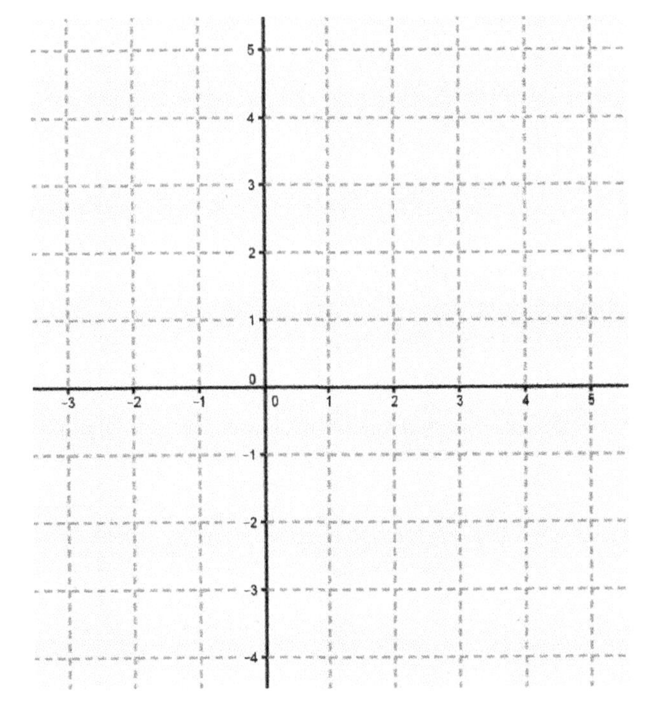

2. Express each of the following in $a + bi$ form.
   a. $(13 + 4i) + (7 + 5i)$
   b. $(5 - i) - 2(1 - 3i)$
   c. $\left((5 - i) - 2(1 - 3i)\right)^2$
   d. $(3 - i)(4 + 7i)$
   e. $(3 - i)(4 + 7i) - \left((5 - i) - 2(1 - 3i)\right)$

3. Express each of the following in $a + bi$ form.
   a. $(2 + 5i) + (4 + 3i)$
   b. $(-1 + 2i) - (4 - 3i)$
   c. $(4 + i) + (2 - i) - (1 - i)$
   d. $(5 + 3i)(3 + 5i)$
   e. $-i(2 - i)(5 + 6i)$
   f. $(1 + i)(2 - 3i) + 3i(1 - i) - i$

4. Find the real values of x and y in each of the following equations using the fact that if $a + bi = c + di$, then $a = c$ and $b = d$.
   a. $5x + 3yi = 20 + 9i$
   b. $2(5x + 9) = (10 - 3y)i$
   c. $3(7 - 2x) - 5(4y - 3)i = x - 2(1 + y)i$

5. Since $i^2 = -1$, we see that

$$i^3 = i^2 \cdot i = -1 \cdot i = -i$$
$$i^4 = i^2 \cdot i^2 = -1 \cdot -1 = 1.$$

Plot i, i^2, i^3, and i^4 on the complex plane, and describe how multiplication by each rotates points in the complex plane.

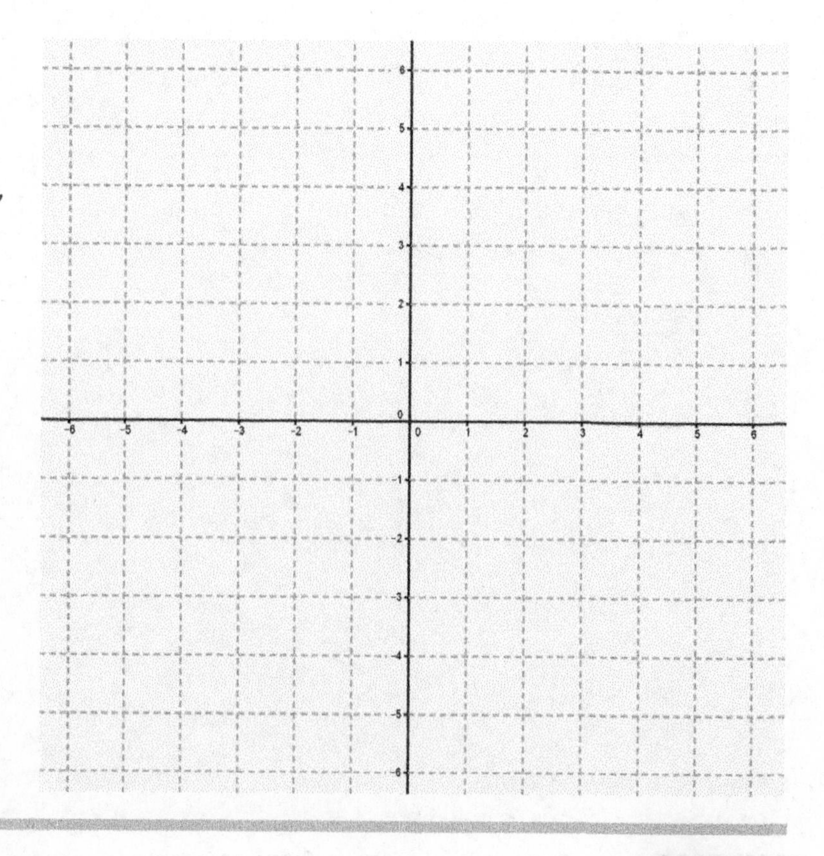

6. Express each of the following in $a + bi$ form.
   a. i^5
   b. i^6
   c. i^7
   d. i^8
   e. i^{102}

7. Express each of the following in $a + bi$ form.
   a. $(1 + i)^2$
   b. $(1 + i)^4$
   c. $(1 + i)^6$

8. Evaluate $x^2 - 6x$ when $x = 3 - i$.

9. Evaluate $4x^2 - 12x$ when $x = \frac{3}{2} - \frac{i}{2}$.

10. Show by substitution that $\frac{5 - i\sqrt{5}}{5}$ is a solution to $5x^2 - 10x + 6 = 0$.

11.
   a. Evaluate the four products below.
      Evaluate $\sqrt{9} \cdot \sqrt{4}$.
      Evaluate $\sqrt{9} \cdot \sqrt{-4}$.
      Evaluate $\sqrt{-9} \cdot \sqrt{4}$.
      Evaluate $\sqrt{-9} \cdot \sqrt{-4}$.
   b. Suppose a and b are positive real numbers. Determine whether the following quantities are equal or not equal.
      $\sqrt{a} \cdot \sqrt{b}$ and $\sqrt{-a} \cdot \sqrt{-b}$
      $\sqrt{-a} \cdot \sqrt{b}$ and $\sqrt{a} \cdot \sqrt{-b}$

This page intentionally left blank

Lesson 38: Complex Numbers as Solutions to Equations

Classwork

Opening Exercises

1. The expression under the radical in the quadratic formula, $b^2 - 4ac$, is called the *discriminant*.

   Use the quadratic formula to solve the following quadratic equations. Calculate the discriminant for each equation.

   a. $x^2 - 9 = 0$

   b. $x^2 - 6x + 9 = 0$

   c. $x^2 + 9 = 0$

2. How does the value of the discriminant for each equation relate the number of solutions you found?

Example 1

Consider the equation $3x + x^2 = -7$.

What does the value of the discriminant tell us about number of solutions to this equation?

Solve the equation. Does the number of solutions match the information provided by the discriminant? Explain.

Exercise

Compute the value of the discriminant of the quadratic equation in each part. Use the value of the discriminant to predict the number and type of solutions. Find all real and complex solutions.

a. $x^2 + 2x + 1 = 0$

b. $x^2 + 4 = 0$

c. $9x^2 - 4x - 14 = 0$

d. $3x^2 + 4x + 2 = 0$

e. $x = 2x^2 + 5$

f. $8x^2 + 4x + 32 = 0$

Lesson Summary

- A quadratic equation with real coefficients may have real or complex solutions.
- Given a quadratic equation $ax^2 + bx + c = 0$, the discriminant $b^2 - 4ac$ indicates whether the equation has two distinct real solutions, one real solution, or two complex solutions.
  - If $b^2 - 4ac > 0$, there are two real solutions to $ax^2 + bx + c = 0$.
  - If $b^2 - 4ac = 0$, there is one real solution to $ax^2 + bx + c = 0$.
  - If $b^2 - 4ac < 0$, there are two complex solutions to $ax^2 + bx + c = 0$.

Problem Set

1. Give an example of a quadratic equation in standard form that has ...
   a. Exactly two distinct real solutions.
   b. Exactly one distinct real solution.
   c. Exactly two complex (non-real) solutions.

2. Suppose we have a quadratic equation $ax^2 + bx + c = 0$ so that $a + c = 0$. Does the quadratic equation have one solution or two distinct solutions? Are they real or complex? Explain how you know.

3. Solve the equation $5x^2 - 4x + 3 = 0$.

4. Solve the equation $2x^2 + 8x = -9$.

5. Solve the equation $9x - 9x^2 = 3 + x + x^2$.

6. Solve the equation $3x^2 - x + 1 = 0$.

7. Solve the equation $6x^4 + 4x^2 - 3x + 2 = 2x^2(3x^2 - 1)$.

8. Solve the equation $25x^2 + 100x + 200 = 0$.

9. Write a quadratic equation in standard form such that -5 is its only solution.

10. Is it possible that the quadratic equation $ax^2 + bx + c = 0$ has a positive real solution if a, b, and c are all positive real numbers?

11. Is it possible that the quadratic equation $ax^2 + bx + c = 0$ has a positive real solution if a, b, and c are all negative real numbers?

Extension:

12. Show that if $k > 3.2$, the solutions of $5x^2 - 8x + k = 0$ are not real numbers.

13. Let k be a real number, and consider the quadratic equation $(k + 1)x^2 + 4kx + 2 = 0$.
    a. Show that the discriminant of $(k + 1)x^2 + 4kx + 2 = 0$ defines a quadratic function of k.
    b. Find the zeros of the function in part (a), and make a sketch of its graph.
    c. For what value of k are there two distinct real solutions to the original quadratic equation?
    d. For what value of k are there two complex solutions to the given quadratic equation?
    e. For what value of k is there one solution to the given quadratic equation?

14. We can develop two formulas that can help us find errors in calculated solutions of quadratic equations.
    a. Find a formula for the sum S of the solutions of the quadratic equation $ax^2 + bx + c = 0$.
    b. Find a formula for the product R of the solutions of the quadratic equation $ax^2 + bx + c = 0$.
    c. June calculated the solutions 7 and -1 to the quadratic equation $x^2 - 6x + 7 = 0$. Do the formulas from parts (a) and (b) detect an error in her solutions? If not, determine if her solution is correct.
    d. Paul calculated the solutions $3 - i\sqrt{2}$ and $3 + i\sqrt{2}$ to the quadratic equation $x^2 - 6x + 7 = 0$. Do the formulas from parts (a) and (b) detect an error in his solutions? If not, determine if his solutions are correct.
    e. Joy calculated the solutions $3 - \sqrt{2}$ and $3 + \sqrt{2}$ to the quadratic equation $x^2 - 6x + 7 = 0$. Do the formulas from parts (a) and (b) detect an error in her solutions? If not, determine if her solutions are correct.
    f. If you find solutions to a quadratic equation that match the results from parts (a) and (b), does that mean your solutions are correct?
    g. Summarize the results of this exercise.

Lesson 39: Factoring Extended to the Complex Realm

Classwork

Opening Exercise

Rewrite each expression as a polynomial in standard form.

a. $(x + i)(x - i)$

b. $(x + 5i)(x - 5i)$

c. $\big(x - (2 + i)\big)\big(x - (2 - i)\big)$

Exercises 1–4

Factor the following polynomial expressions into products of linear terms.

1. $x^2 + 9$

2. $x^2 + 5$

3. Consider the polynomial $P(x) = x^4 - 3x^2 - 4$.

   a. What are the solutions to $x^4 - 3x^2 - 4 = 0$?

   b. How many x-intercepts does the graph of the equation $y = x^4 - 3x^2 - 4$ have? What are the coordinates of the x-intercepts?

   c. Are solutions to the polynomial equation $P(x) = 0$ the same as the x-intercepts of the graph of $y = P(x)$? Justify your reasoning.

4. Write a polynomial P with the lowest possible degree that has the given solutions. Explain how you generated each answer.

   a. $-2, 3, -4i, 4i$

   b. $-1, 3i$

   c. $0, 2, 1+i, 1-i$

   d. $\sqrt{2}, -\sqrt{2}, 3, 1+2i$

   e. $2i, 3-i$

Lesson Summary

- Polynomial equations with real coefficients can have real or complex solutions or they can have both.
- If a complex number is a solution to a polynomial equation, then its conjugate is also a solution.
- Real solutions to polynomial equations correspond to the x-intercepts of the associated graph, but complex solutions do not.

Problem Set

1. Rewrite each expression in standard form.
   a. $(x + 3i)(x - 3i)$
   b. $(x - a + bi)(x - (a + bi))$
   c. $(x + 2i)(x - i)(x + i)(x - 2i)$
   d. $(x + i)^2 \cdot (x - i)^2$

2. Suppose in Problem 1 that you had no access to paper, writing utensils, or technology. How do you know that the expressions in parts (a)–(d) are polynomials with real coefficients?

3. Write a polynomial equation of degree 4 in standard form that has the solutions $i, -i, 1, -1$.

4. Explain the difference between x-intercepts and solutions to an equation. Give an example of a polynomial with real coefficients that has twice as many solutions as x-intercepts. Write it in standard form.

5. Find the solutions to $x^4 - 5x^2 - 36 = 0$ and the x-intercepts of the graph of $y = x^4 - 5x^2 - 36$.

6. Find the solutions to $2x^4 - 24x^2 + 40 = 0$ and the x-intercepts of the graph of $y = 2x^4 - 24x^2 + 40$.

7. Find the solutions to $x^4 - 64 = 0$ and the x-intercepts of the graph of $y = x^4 - 64$.

8. Use the fact that $x^4 + 64 = (x^2 - 4x + 8)(x^2 + 4x + 8)$ to explain how you know that the graph of $y = x^4 + 64$ has no x-intercepts. You need not find the solutions.

Lesson 40: Obstacles Resolved—A Surprising Result

Classwork

Opening Exercise

Write each of the following quadratic expressions as a product of linear factors. Verify that the factored form is equivalent.

a. $x^2 + 12x + 27$

b. $x^2 - 16$

c. $x^2 + 16$

d. $x^2 + 4x + 5$

Example 1

Consider the polynomial $P(x) = x^3 + 3x^2 + x - 5$ whose graph is shown to the right.

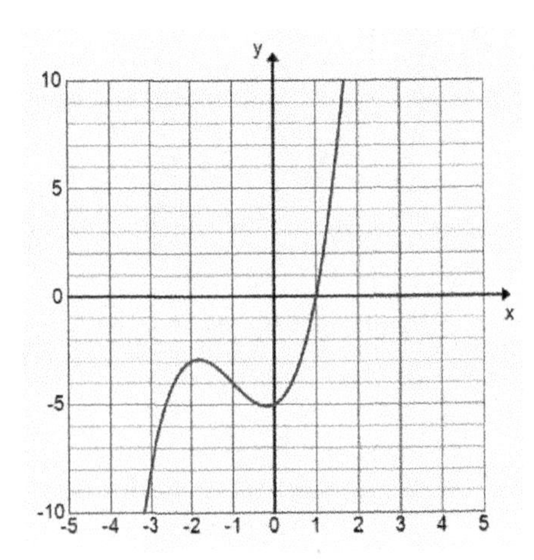

a. Looking at the graph, how do we know that there is only one real solution?

b. Is it possible for a cubic polynomial function to have no zeros?

c. From the graph, what appears to be one solution to the equation $x^3 + 3x^2 + x - 5 = 0$?

d. How can we verify that this value is a solution?

e. According to the remainder theorem, what is one factor of the cubic expression $x^3 + 3x^2 + x - 5$?

f. Factor out the expression you found in part (e) from $x^3 + 3x^2 + x - 5$.

g. What are all of the solutions to $x^3 + 3x^2 + x - 5 = 0$?

h. Write the expression $x^3 + 3x^2 + x - 5$ in terms of linear factors.

Exercises 1–2

Write each polynomial in terms of linear factors. The graph of $y = x^3 - 3x^2 + 4x - 12$ is provided for Exercise 2.

1. $f(x) = x^3 + 5x$

2. $g(x) = x^3 - 3x^2 + 4x - 12$

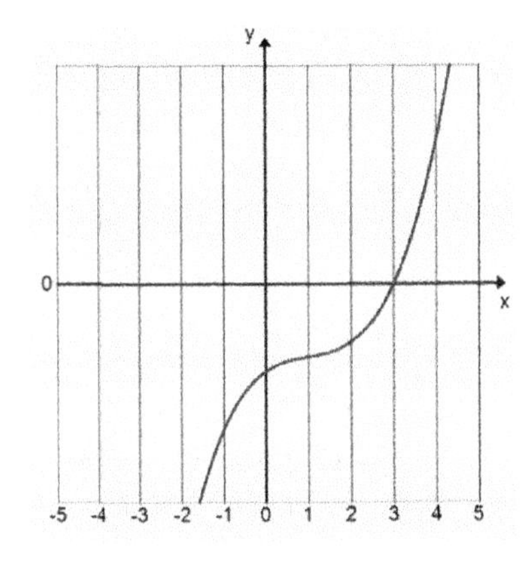

Example 2

Consider the polynomial function $P(x) = x^4 - 3x^3 + 6x^2 - 12x + 8$, whose corresponding graph $y = x^4 - 3x^3 + 6x^2 - 12x + 8$ is shown to the right. How many zeros does P have?

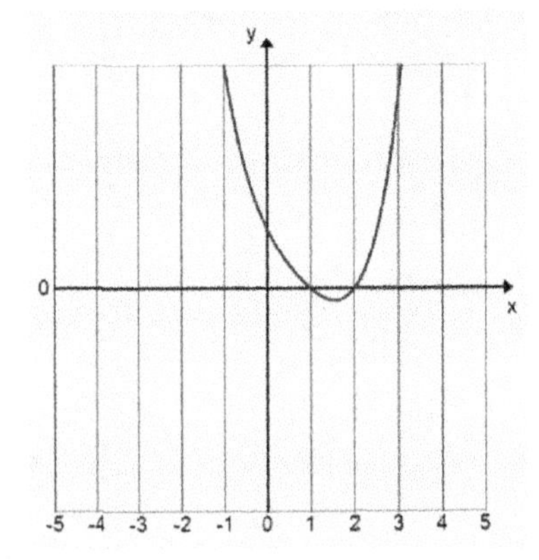

a. Part 1 of the fundamental theorem of algebra says that this equation will have at least one solution in the complex numbers. How does this align with what we can see in the graph to the right?

b. Identify one zero from the graph.

c. Use polynomial division to factor out one linear term from the expression $x^4 - 3x^3 + 6x^2 - 12x + 8$.

d. Now we have a cubic polynomial to factor. We know by part 1 of the fundamental theorem of algebra that a polynomial function will have at least one real zero. What is that zero in this case?

e. Use polynomial division to factor out another linear term of $x^4 - 3x^3 + 6x^2 - 12x + 8$.

f. Are we done? Can we factor this polynomial any further?

g. Now that the polynomial is in factored form, we can quickly see how many solutions there are to the original equation $x^4 - 3x^3 + 6x^2 - 12x + 8 = 0$.

h. What if we had started with a polynomial function of degree 8?

Lesson Summary

Every polynomial function of degree n, for $n \geq 1$, has n roots over the complex numbers, counted with multiplicity. Therefore, such polynomials can always be factored into n linear factors, and the obstacles to factoring we saw before have all disappeared in the larger context of allowing solutions to be complex numbers.

The Fundamental Theorem of Algebra:

1. If P is a polynomial function of degree $n \geq 1$, with real or complex coefficients, then there exists at least one number r (real or complex) such that $P(r) = 0$.
2. If P is a polynomial function of degree $n \geq 1$, given by $P(x) = a_n x^n + a_{n-1}x^{n-1} + \cdots + a_1 x + a_0$ with real or complex coefficients a_i, then P has exactly n zeros $r_1, r_2, \ldots, r_n$ (not all necessarily distinct), such that $P(x) = a_n(x - r_1)(x - r_2)\cdots(x - r_n)$.

Problem Set

1. Write each quadratic function below in terms of linear factors.
   a. $f(x) = x^2 - 25$
   b. $f(x) = x^2 + 25$
   c. $f(x) = 4x^2 + 25$
   d. $f(x) = x^2 - 2x + 1$
   e. $f(x) = x^2 - 2x + 4$

2. Consider the polynomial function $P(x) = (x^2 + 4)(x^2 + 1)(2x + 3)(3x - 4)$.
   a. Express P in terms of linear factors.
   b. Fill in the blanks of the following sentence.

   The polynomial P has degree ________________ and can, therefore, be written in terms of _________ linear factors.

   The function P has __________ zeros. There are ____________ real zeros and _____ complex zeros. The graph of $y = P(x)$ has _______ x-intercepts.

3. Express each cubic function below in terms of linear factors.
   a. $f(x) = x^3 - 6x^2 - 27x$
   b. $f(x) = x^3 - 16x^2$
   c. $f(x) = x^3 + 16x$

4. For each cubic function below, one of the zeros is given. Express each cubic function in terms of linear factors.
   a. $f(x) = 2x^3 - 9x^2 - 53x - 24; f(8) = 0$
   b. $f(x) = x^3 + x^2 + 6x + 6; f(-1) = 0$

5. Determine if each statement is always true or sometimes false. If it is sometimes false, explain why it is not always true.
   a. A degree 2 polynomial function will have two linear factors.
   b. The graph of a degree 2 polynomial function will have two x-intercepts.
   c. The graph of a degree 3 polynomial function might not cross the x-axis.
   d. A polynomial function of degree n can be written in terms of n linear factors.

6. Consider the polynomial function $f(x) = x^6 - 9x^3 + 8$.
   a. How many linear factors does $x^6 - 9x^3 + 8$ have? Explain.
   b. How is this information useful for finding the zeros of f?
   c. Find the zeros of f. (Hint: Let $Q = x^3$. Rewrite the equation in terms of Q to factor.)

7. Consider the polynomial function $P(x) = x^4 - 6x^3 + 11x^2 - 18$.
   a. Use the graph to find the real zeros of P.
   b. Confirm that the zeros are correct by evaluating the function P at those values.
   c. Express P in terms of linear factors.
   d. Find all zeros of P.

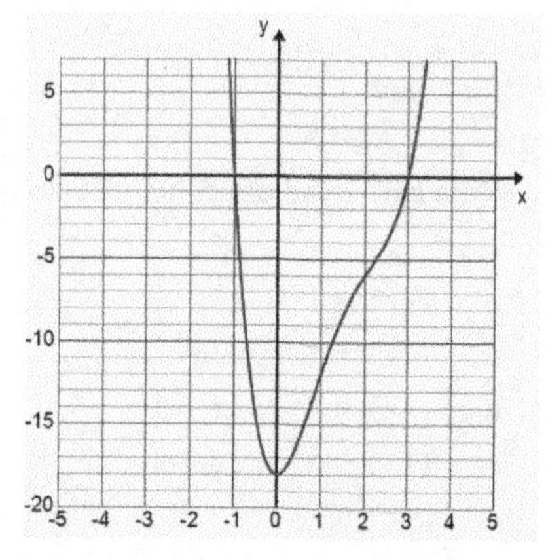

8. Penny says that the equation $x^3 - 8 = 0$ has only one solution, $x = 2$. Use the fundamental theorem of algebra to explain to her why she is incorrect.

9. Roger says that the equation $x^2 - 12x + 36 = 0$ has only one solution, 6. Regina says Roger is wrong and that the fundamental theorem of algebra guarantees that a quadratic equation must have two solutions. Who is correct and why?

Student Edition

Eureka Math

Algebra II
Module 2

Special thanks go to the Gordan A. Cain Center and to the Department of Mathematics at Louisiana State University for their support in the development of *Eureka Math*.

Published by Great Minds

Printed in the U.S.A.
This book may be purchased from the publisher at eureka-math.org
10 9 8 7 6 5 4 3

ISBN 978-1-63255-330-0

Lesson 1: Ferris Wheels—Tracking the Height of a Passenger Car

Classwork

Exploratory Challenge 1: The Height of a Ferris Wheel Car

George Ferris built the first Ferris wheel in 1893 for the World's Columbian Exhibition in Chicago. It had 30 passenger cars, was 264 feet tall and rotated once every 9 minutes when all the cars were loaded. The ride cost $0.50.

Source: The New York Times/Redux

a. Create a sketch of the height of a passenger car on the original Ferris wheel as that car rotates around the wheel 4 times. List any assumptions that you are making as you create your model.

b. What type of function would best model this situation?

Exercises 1–5

1. Suppose a Ferris wheel has a diameter of 150 feet. From your viewpoint, the Ferris wheel is rotating counterclockwise. We will refer to a rotation through a full $360°$ as a *turn*.

   a. Create a sketch of the height of a car that starts at the bottom of the wheel and continues for two turns.

   b. Explain how the features of your graph relate to this situation.

2. Suppose a Ferris wheel has a diameter of 150 feet. From your viewpoint, the Ferris wheel is rotating counterclockwise.

   a. Your friends board the Ferris wheel, and the ride continues boarding passengers. Their car is in the three o'clock position when the ride begins. Create a sketch of the height of your friends' car for two turns.

b. Explain how the features of your graph relate to this situation.

3. How would your sketch change if the diameter of the wheel changed?

4. If you translated the sketch of your graph down by the radius of the wheel, what would the x-axis represent in this situation?

5. How could we create a more precise sketch?

Exploratory Challenge 2: The Paper Plate Model

Use a paper plate mounted on a sheet of paper to model a Ferris wheel, where the lower edge of the paper represents the ground. Use a ruler and protractor to measure the height of a Ferris wheel car above the ground for various amounts of rotation. Suppose that your friends board the Ferris wheel near the end of the boarding period and the ride begins when their car is in the three o'clock position as shown.

a. Mark the diagram below to estimate the location of the Ferris wheel passenger car every 15 degrees. The point on the circle below represents the passenger car in the 3 o'clock position. Since this is the beginning of the ride, consider this position to be the result of rotating by $0°$.

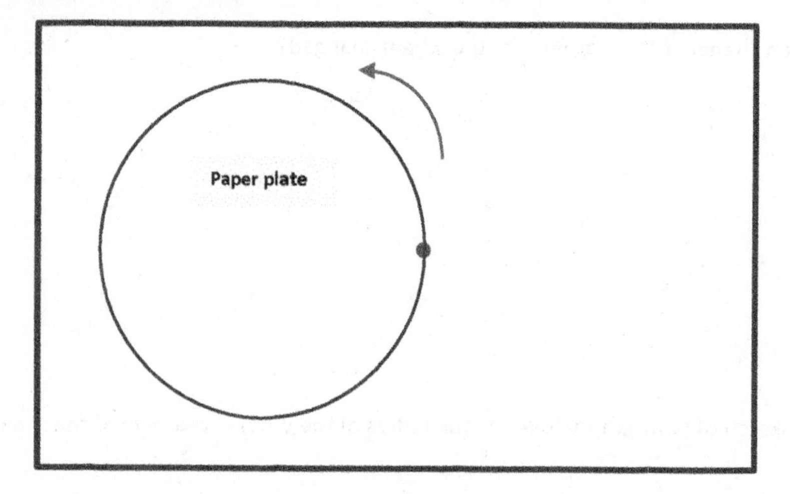

b. Using the physical model you created with your group, record your measurements in the table, and then graph the ordered pairs (rotation, height) on the coordinate grid shown below. Provide appropriate labels on the axes.

Rotation (degrees)	Height (cm)
0	
15	
30	
45	
60	
75	
90	

Rotation (degrees)	Height (cm)
105	
120	
135	
150	
165	
180	
195	

Rotation (degrees)	Height (cm)
210	
225	
240	
255	
270	
285	
300	

Rotation (degrees)	Height (cm)
315	
330	
345	
360	

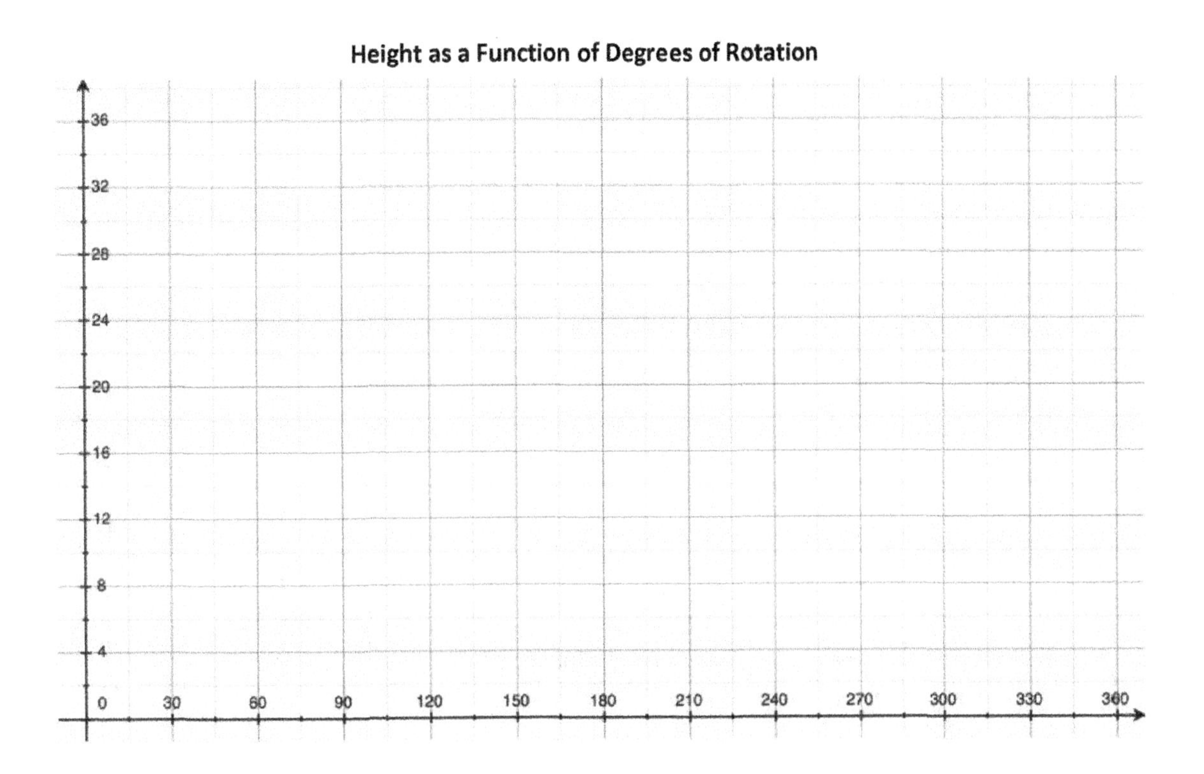

c. Explain how the features of your graph relate to the paper plate model you created.

Closing

- How does a function like the one that represents the height of a passenger car on a Ferris wheel differ from other types of functions you have studied such as linear, polynomial, and exponential functions?

- What is the domain of your Ferris wheel height function? What is the range?

- Provide a definition of periodic function in your own words. Why is the Ferris wheel height function an example of a periodic function?

- What other situations might be modeled by a periodic function?

Problem Set

1. Suppose that a Ferris wheel is 40 feet in diameter and rotates counterclockwise. When a passenger car is at the bottom of the wheel, it is located 2 feet above the ground.
   a. Sketch a graph of a function that represents the height of a passenger car that starts at the 3 o'clock position on the wheel for one turn.
   b. Sketch a graph of a function that represents the height of a passenger car that starts at the top of the wheel for one turn.
   c. The sketch you created in part (a) represents a graph of a function. What is the domain of the function? What is the range?
   d. The sketch you created in part (b) represents a graph of a function. What is the domain of the function? What is the range?
   e. Describe how the graph of the function in part (a) would change if you sketched the graph for two turns.
   f. Describe how the function in part (a) and its graph would change if the Ferris wheel had a diameter of 60 feet.

2. A small pebble is lodged in the tread of a tire with radius 25 cm. Sketch the height of the pebble above the ground as the tire rotates counterclockwise through 5 turns. Start your graph when the pebble is at the 9 o'clock position.

3. The graph you created in Problem 2 represents a function.
   a. Describe how the function and its graph would change if the tire's radius was 24 inches instead of 25 cm.
   b. Describe how the function and its graph would change if the wheel was turning in the opposite direction.
   c. Describe how the function and its graph would change if we started the graph when the pebble was at ground level.

4. Justice believes that the height of a Ferris wheel passenger car is best modeled with a piecewise linear function. Make a convincing argument why a piecewise linear function is NOT a good model for the height of a car on a rotating Ferris wheel.

This page intentionally left blank

Lesson 2: The Height and Co-Height Functions of a Ferris Wheel

Classwork

Opening Exercise

Suppose a Ferris wheel has a radius of 50 feet. We will measure the height of a passenger car that starts in the 3 o'clock position with respect to the horizontal line through the center of the wheel. That is, we consider the height of the passenger car at the outset of the problem (that is, after a $0°$ rotation) to be 0 feet.

a. Mark the diagram to show the position of a passenger car at 30-degree intervals as it rotates counterclockwise around the Ferris wheel.

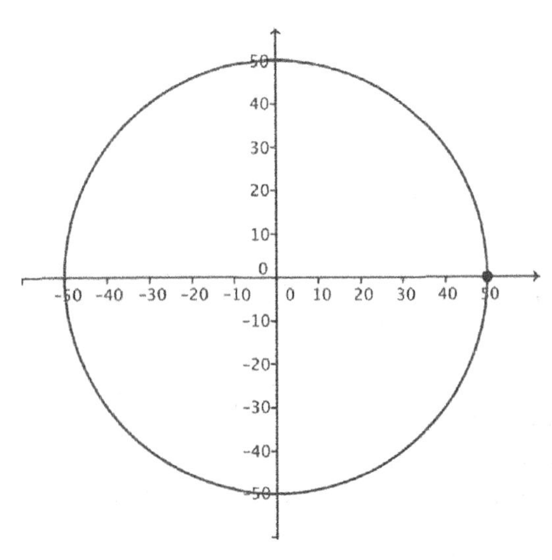

b. Sketch the graph of the height function of the passenger car for one turn of the wheel. Provide appropriate labels on the axes.

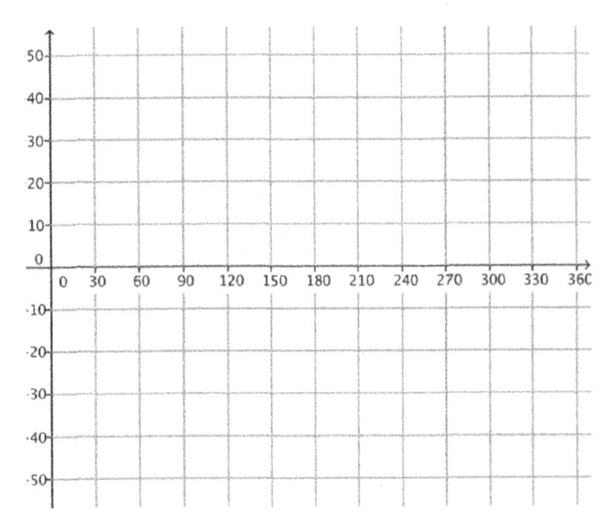

c. Explain how you can identify the radius of the wheel from the graph in part (b).

d. If the center of the wheel is 55 feet above the ground, how high is the passenger car above the ground when it is at the top of the wheel?

Exercises 1–3

1. Each point P_1, P_2, ... P_8 on the circle in the diagram to the right represents a passenger car on a Ferris wheel.

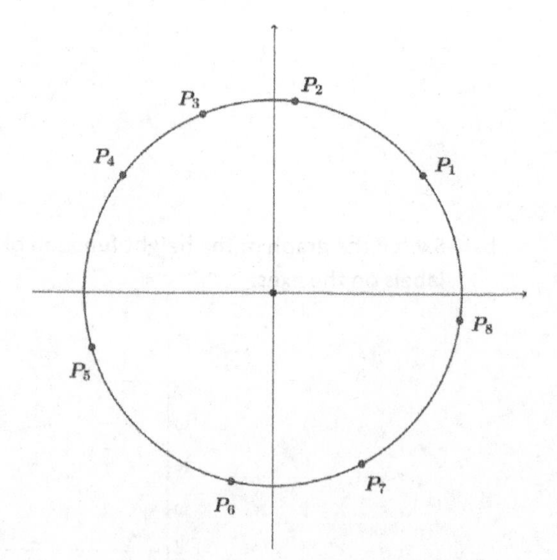

a. Draw segments that represent the co-height of each car. Which cars have a positive co-height? Which cars have a negative co-height?

b. List the points in order of increasing co-height; that is, list the point with the smallest co-height first and the point with the largest co-height last.

2. Suppose that the radius of a Ferris wheel is 100 feet and the wheel rotates counterclockwise through one turn. Define a function that measures the co-height of a passenger car as a function of the degrees of rotation from the initial 3 o'clock position.

   a. What is the domain of the co-height function?

   b. What is the range of the co-height function?

   c. How does changing the wheel's radius affect the domain and range of the co-height function?

3. For a Ferris wheel of radius 100 feet going through one turn, how do the domain and range of the height function compare to the domain and range of the co-height function? Is this true for any Ferris wheel?

Exploratory Challenge: The Paper Plate Model, Revisited

Use a paper plate mounted on a sheet of paper to model a Ferris wheel, where the lower edge of the paper represents the ground. Use a ruler and protractor to measure the height and co-height of a Ferris wheel car at various amounts of rotation, measured with respect to the horizontal and vertical lines through the center of the wheel. Suppose that your friends board the Ferris wheel near the end of the boarding period, and the ride begins when their car is in the three o'clock position as shown.

a. Mark horizontal and vertical lines through the center of the wheel on the card stock behind the plate as shown. We will measure the height and co-height as the displacement from the horizontal and vertical lines through the center of the plate.

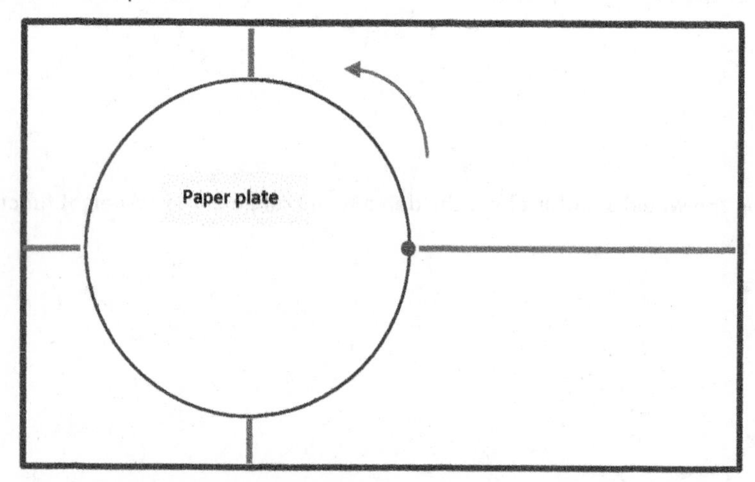

b. Using the physical model you created with your group, record your measurements in the table, and then graph each of the two sets of ordered pairs (rotation angle, height) and (rotation angle, co-height) on separate coordinate grids. Provide appropriate labels on the axes.

Rotation (degrees)	Height (cm)	Co-Height (cm)
0		
15		
30		
45		
60		
75		
90		
105		
120		

Rotation (degrees)	Height (cm)	Co-Height (cm)
135		
150		
165		
180		
195		
210		
225		
240		

Rotation (degrees)	Height (cm)	Co-Height (cm)
255		
270		
285		
300		
315		
330		
345		
360		

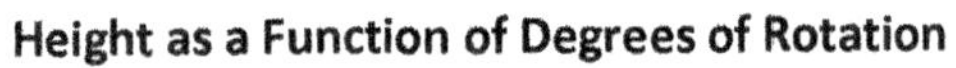

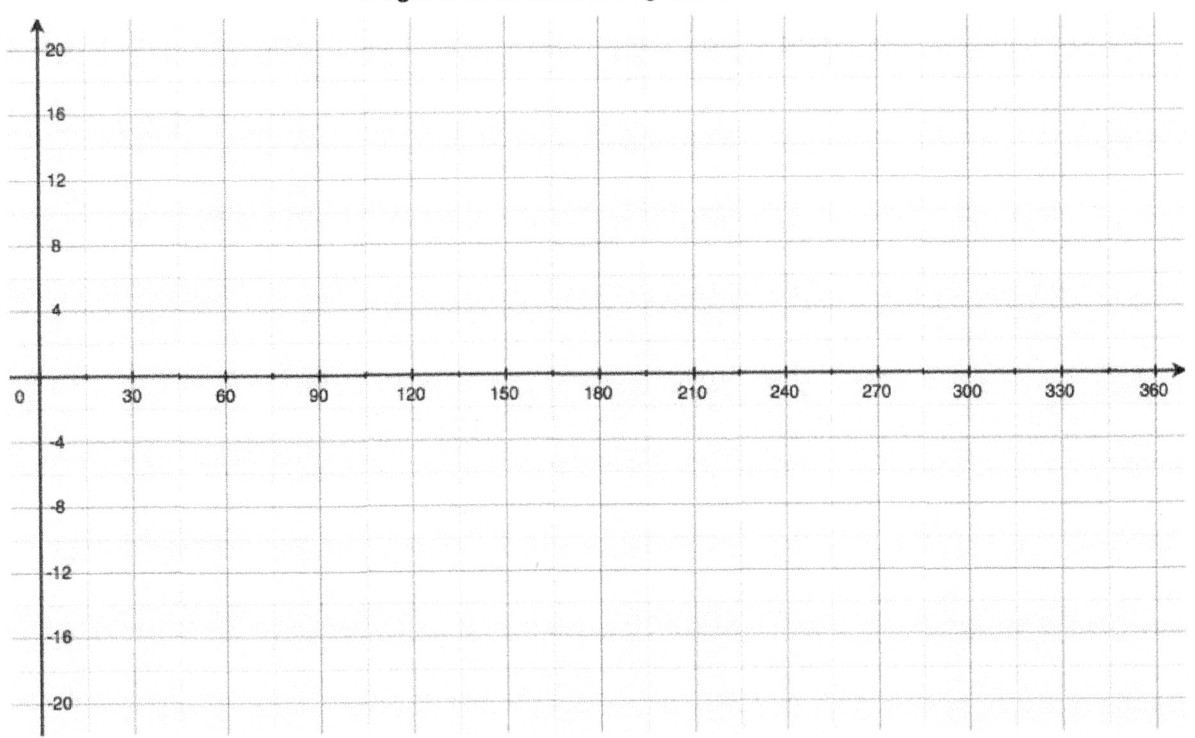

Co-Height as a Function of Degrees of Rotation

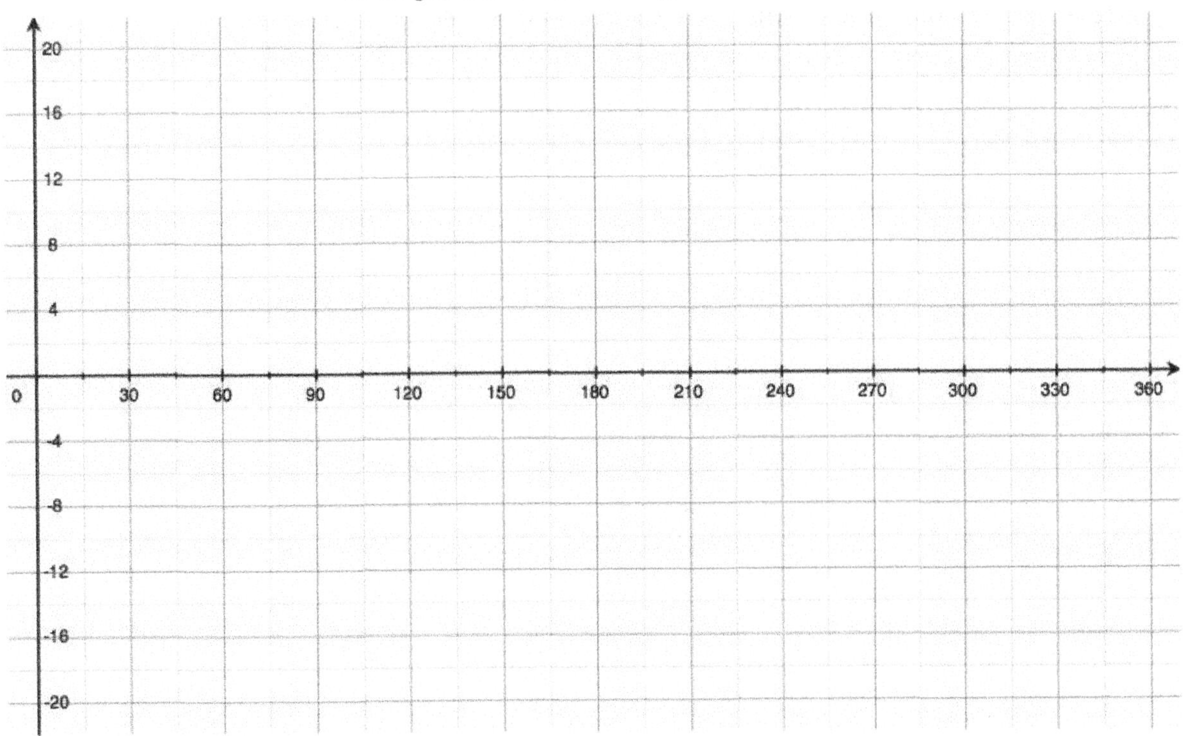

Closing

- Why do you think we named the new function the co-height?

- How are the graphs of these two functions alike? How are they different?

- What does a negative value of the height function tell us about the location of the passenger car at various positions around a Ferris wheel? What about a negative value of the co-height function?

Problem Set

1. The Seattle Great Wheel, with an overall height of 175 feet, was the tallest Ferris wheel on the West Coast at the time of its construction in 2012. For this exercise, assume that the diameter of the wheel is 175 feet.

   a. Create a diagram that shows the position of a passenger car on the Great Wheel as it rotates counterclockwise at 45-degree intervals.
   b. On the same set of axes, sketch graphs of the height and co-height functions for a passenger car starting at the 3 o'clock position on the Great Wheel and completing one turn.
   c. Discuss the similarities and differences between the graph of the height function and the graph of the co-height function.
   d. Explain how you can identify the radius of the wheel from either graph.

2. In 2014, the High Roller Ferris wheel opened in Las Vegas, dwarfing the Seattle Great Wheel with a diameter of 520 feet. Sketch graphs of the height and co-height functions for one complete turn of the High Roller.

3. Consider a Ferris wheel with a 50-foot radius. We will track the height and co-height of passenger cars that begin at the 3 o'clock position. Sketch graphs of the height and co-height functions for the following scenarios.
   a. A passenger car on the Ferris wheel completes one turn, traveling counterclockwise.
   b. A passenger car on the Ferris wheel completes two full turns, traveling counterclockwise.
   c. The Ferris wheel is stuck in reverse, and a passenger car on the Ferris wheel completes two full *clockwise* turns.

4. Consider a Ferris wheel with radius of 40 feet that is rotating counterclockwise. At which amounts of rotation are the values of the height and co-height functions equal? Does this result hold for a Ferris wheel with a different radius?

5. Yuki is on a passenger car of a Ferris wheel at the 3 o'clock position. The wheel then rotates 135 degrees counterclockwise and gets stuck. Lee argues that she can compute the value of the co-height of Yuki's car if she is given one of the following two pieces of information:
   i. The value of the height function of Yuki's car, or
   ii. The diameter of the Ferris wheel itself.

   Is Lee correct? Explain how you know.

This page intentionally left blank

Lesson 3: The Motion of the Moon, Sun, and Stars—Motivating Mathematics

Classwork

Opening

Why does it look like the sun moves across the sky?

Is the sun moving, or are you moving?

In ancient Greek mythology, the god Helios was the personification of the sun. He rode across the sky every day in his chariot led by four horses. Why do your answers make it believable that in ancient times people imagined the sun was pulled across the sky each day?

Discussion

In mathematics, counterclockwise rotation is considered to be the positive direction of rotation, which runs counter to our experience with a very common example of rotation: the rotation of the hands on a clock.

- Is there a connection between counterclockwise motion being considered to be positive and the naming of the quadrants on a standard coordinate system?

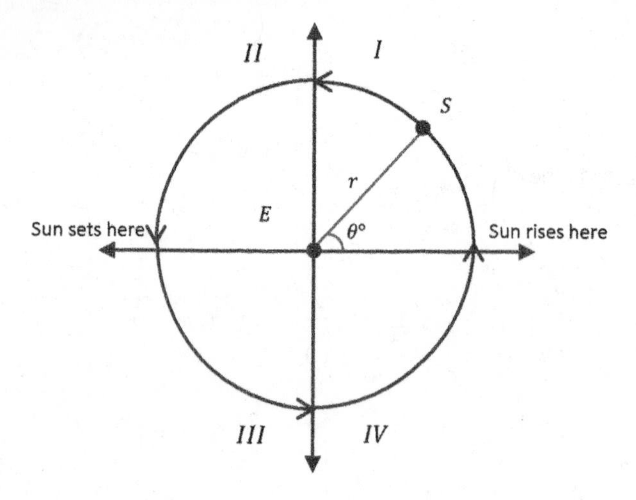

- What does the circle's radius, r, represent?

- How has the motion of the sun influenced the development of mathematics?

- How is measuring the *height* of the sun like measuring the Ferris wheel passenger car height in the previous lessons?

Exercises 1–4

1. Calculate jya($7°$), jya($11°$), jya($15°$), and jya($18°$) using Aryabhata's formula[1], round to the nearest integer, and add your results to the table below. Leave the rightmost column blank for now.

n	θ, in degrees	jya($\theta°$)	$3438\sin(\theta°)$
1	$3\frac{3}{4}$	225	
2	$7\frac{1}{2}$		
3	$11\frac{1}{4}$		
4	15		
5	$18\frac{3}{4}$		
6	$22\frac{1}{2}$	1315	
7	$26\frac{1}{4}$	1520	
8	30	1719	
9	$33\frac{3}{4}$	1910	
10	$37\frac{1}{2}$	2093	
11	$41\frac{1}{4}$	2267	
12	45	2431	

n	θ, in degrees	jya($\theta°$)	$3438\sin(\theta°)$
13	$48\frac{3}{4}$	2585	
14	$52\frac{1}{2}$	2728	
15	$56\frac{1}{4}$	2859	
16	60	2978	
17	$63\frac{3}{4}$	3084	
18	$67\frac{1}{2}$	3177	
19	$71\frac{1}{4}$	3256	
20	75	3321	
21	$78\frac{3}{4}$	3372	
22	$82\frac{1}{2}$	3409	
23	$86\frac{1}{4}$	3431	
24	90	3438	

[1] In constructing the table, Aryabhata made adjustments to the values of his approximation to the jya to match his observational data. The first adjustment occurs in the calculation of jya($30°$). Thus, the entire table cannot be accurately constructed using this formula.

2. Label the angle θ, $\text{jya}(\theta°)$, $\text{kojya}(\theta°)$, and r in the diagram shown below.

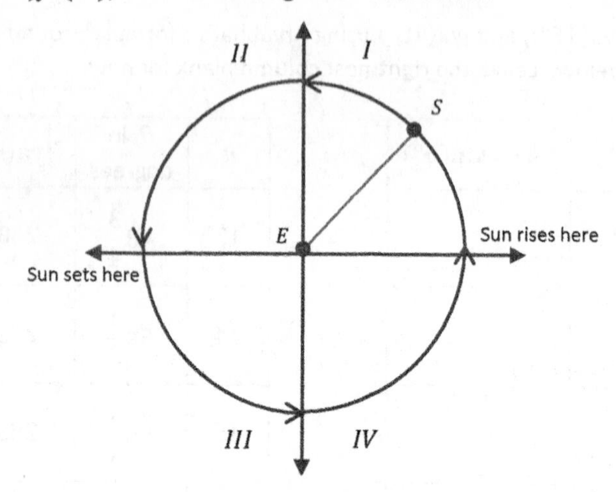

a. How does this relate to something you have done before?

b. How do $\text{jya}(\theta°)$ and $\text{kojya}(\theta°)$ relate to lengths we already know?

3. Use your calculator to compute $r\sin(\theta°)$ for each value of θ in the table from Exercise 1, where $r = 3438$. Record this in the blank column on the right in Exercise 1, rounding to the nearest integer. How do Aryabhata's approximated values from around the year 500 C.E. compare to the value we can calculate with our modern technology?

4. We will assume that the sun rises at 6:00 a.m., is directly overhead at 12:00 noon, and sets at 6:00 p.m. We measure the *height* of the sun by finding its vertical distance from the horizon line; the horizontal line that connects the easternmost point, where the sun rises, to the westernmost point, where the sun sets.

   a. Using $r = 3438$, as Aryabhata did, find the *height* of the sun at the times listed in the following table:

Time of day	Height
6:00 a.m.	
7:00 a.m.	
8:00 a.m.	
9:00 a.m.	
10:00 a.m.	
11:00 a.m.	
12:00 p.m.	

   b. Now, find the height of the sun at the times listed in the following table using the actual distance from the earth to the sun, which is 93 million miles.

Time of day	Height
6:00 a.m.	
7:00 a.m.	
8:00 a.m.	
9:00 a.m.	
10:00 a.m.	
11:00 a.m.	
12:00 p.m.	

Lesson Summary

Ancient scholars in Babylon and India conjectured that celestial motion was circular; the sun and other stars orbited the earth in a circular fashion. The earth was presumed to be the center of the sun's orbit.

The quadrant numbering in a coordinate system is consistent with the counterclockwise motion of the sun, which rises from the east and sets in the west.

The 6th century Indian scholar Aryabhata created the first sine table, using a measurement he called jya. The purpose of his table was to calculate the position of the sun, the stars, and the planets.

Problem Set

1. An Indian astronomer noted that the angle of his line of sight to Venus measured $52°$. We now know that the average distance from Earth to Venus is 162 million miles. Use Aryabhata's table to estimate the apparent height of Venus. Round your answer to the nearest million miles.

2. Later, the Indian astronomer saw that the angle of his line of sight to Mars measured $82°$. We now know that the average distance from Earth to Mars is 140 million miles. Use Aryabhata's table to estimate the apparent height of Mars. Round your answer to the nearest million miles.

3. The moon orbits the earth in an elongated orbit, with an average distance of the moon from the earth of roughly $239{,}000$ miles. It takes the moon 27.32 days to travel around the earth, so the moon moves with respect to the stars roughly $0.5°$ every hour. Suppose that angle of inclination of the moon with respect to the observer measures $45°$ at midnight. As in Example 1, an observer is standing still and facing north. Use Aryabhata's jya table to find the apparent height of the moon above the observer at the times listed in the table below, to the nearest thousand miles.

Time (hour:min)	Angle of elevation θ, in degrees	Height
12:00 a.m.		
7:30 a.m.		
3:00 p.m.		
10:30 p.m.		
6:00 a.m.		
1:30 p.m.		
9:00 p.m.		

4. George wants to apply Aryabhata's method to estimate the height of the International Space Station, which orbits Earth at a speed of about 17,500 miles per hour. This means that the space station makes one full rotation around Earth roughly every 90 minutes. The space station maintains a low earth orbit, with an average distance from Earth of 238 miles.

   a. George supposes that the space station is just visible on the eastern horizon at 12:00 midnight, so its apparent height at that time would be 0 miles above the horizon. Use Aryabhata's jya table to find the apparent height of the space station above the observer at the times listed in the table below.

Time (hour:min:sec)	Angle of elevation θ, in degrees	Height
12:00:00 a.m.		
12:03:45 a.m.		
12:07:30 a.m.		
12:11:15 a.m.		
12:15:00 a.m.		
12:18:45 a.m.		
12:22:30 a.m.		

   b. When George presents his solution to his classmate Jane, she tells him that his model isn't appropriate for this situation. Is she correct? Explain how you know. (Hint: As we set up our model in the first discussion, we treated our observer as if he was the center of the orbit of the sun around the earth. In part (a) of this problem, we treated our observer as if she were the center of the orbit of the International Space Station around Earth. The radius of Earth is approximately 3963 miles, the space station orbits about 238 miles above Earth's surface, and the distance from Earth to the sun is roughly 93,000,000 miles. Draw a picture of the earth and the path of the space station, and then compare that to the points with heights and rotation angles from part (a).)

This page intentionally left blank

Lesson 4: From Circle-ometry to Trigonometry

Classwork

Opening Exercises

1. Find the lengths of the sides of the right triangles below, each of which has hypotenuse of length 1.

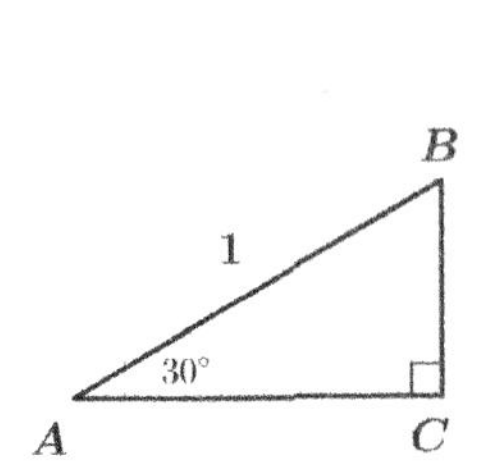

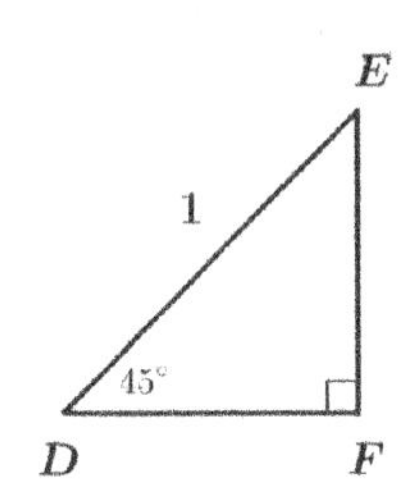

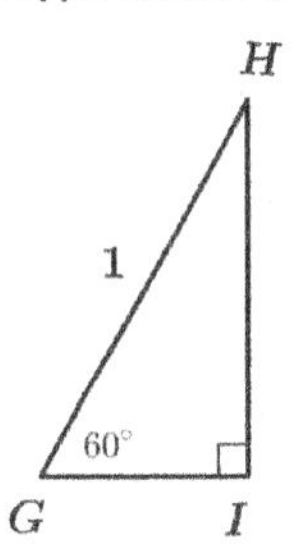

2. Given the following right triangle $\triangle ABC$ with $m\angle A = \theta°$, find $\sin(\theta°)$ and $\cos(\theta°)$.

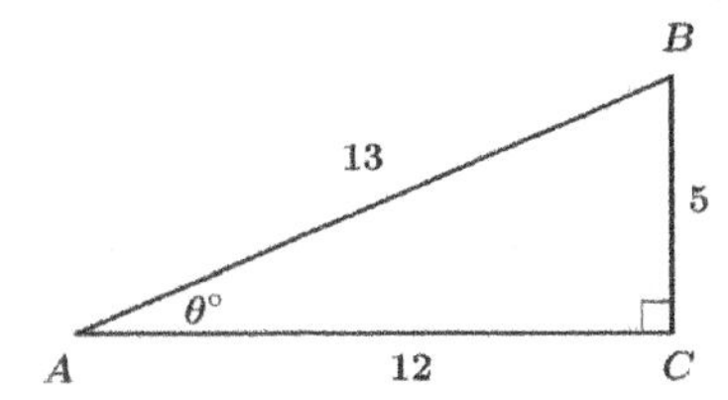

Example 1

Suppose that point P is the point on the unit circle obtained by rotating the initial ray through $30°$. Find $\sin(30°)$ and $\cos(30°)$.

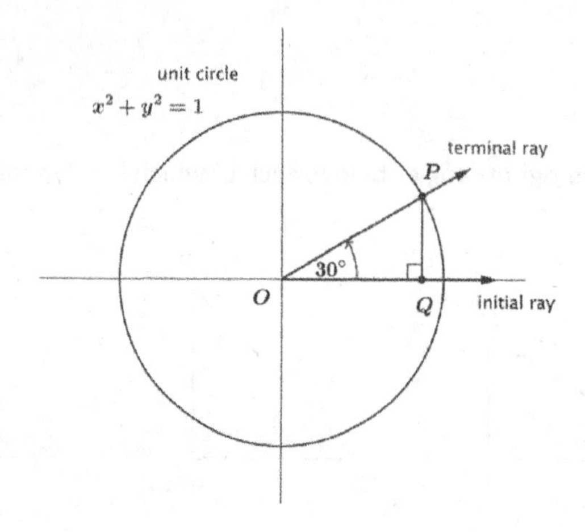

What is the length OQ of the horizontal leg of our triangle?

What is the length QP of the vertical leg of our triangle?

What is $\sin(30°)$?

What is $\cos(30°)$?

Exercises 1–2

1. Suppose that P is the point on the unit circle obtained by rotating the initial ray through $45°$. Find $\sin(45°)$ and $\cos(45°)$.

2. Suppose that P is the point on the unit circle obtained by rotating the initial ray through $60°$. Find $\sin(60°)$ and $\cos(60°)$.

Example 2

Suppose that P is the point on the unit circle obtained by rotating the initial ray through $150°$. Find $\sin(150°)$ and $\cos(150°)$.

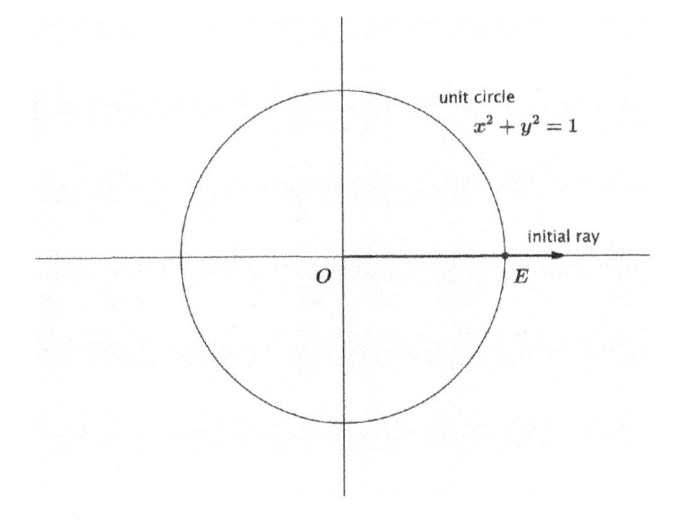

Discussion

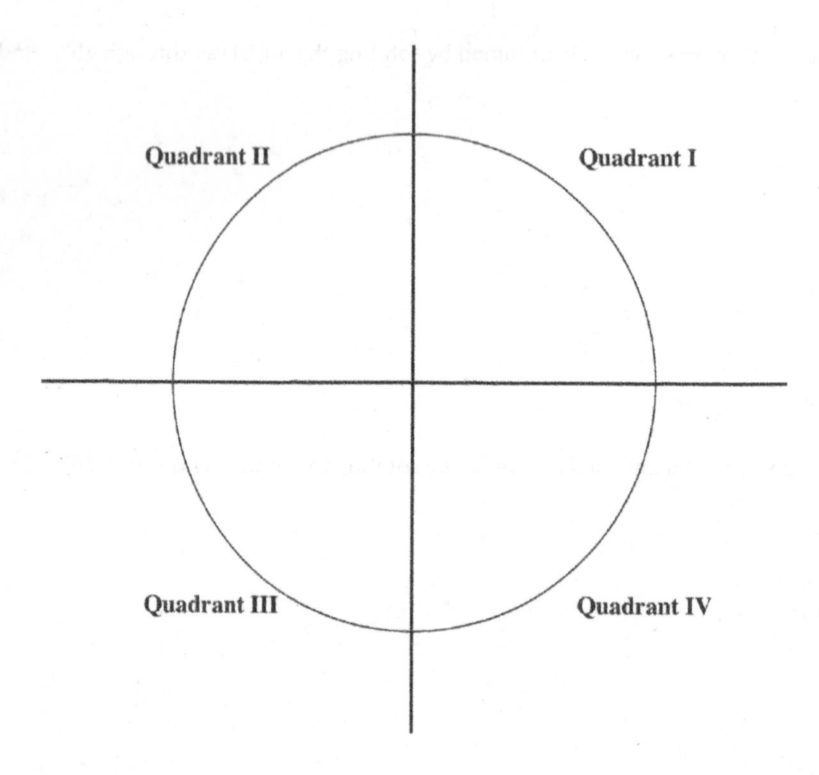

Exercises 3–5

3. Suppose that P is the point on the unit circle obtained by rotating the initial ray counterclockwise through 120 degrees. Find the measure of the reference angle for $120°$, and then find $\sin(120°)$ and $\cos(120°)$.

4. Suppose that P is the point on the unit circle obtained by rotating the initial ray counterclockwise through $240°$. Find the measure of the reference angle for $240°$, and then find $\sin(240°)$ and $\cos(240°)$.

5. Suppose that P is the point on the unit circle obtained by rotating the initial ray counterclockwise through 330 degrees. Find the measure of the reference angle for $330°$, and then find $\sin(330°)$ and $\cos(330°)$.

Discussion

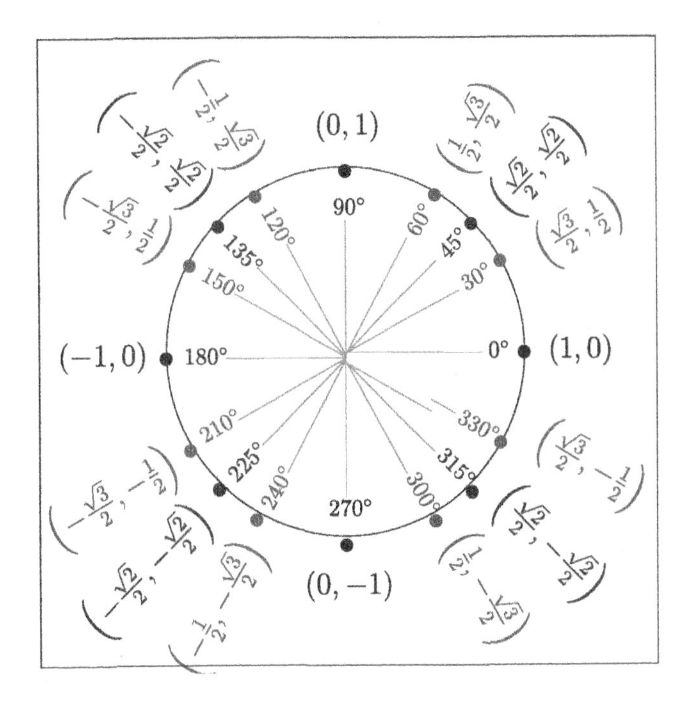

Lesson Summary

In this lesson we formalized the idea of the height and co-height of a Ferris wheel and defined the sine and cosine functions that give the x- and y- coordinates of the intersection of the unit circle and the initial ray rotated through θ degrees, for most values of θ with $0 < \theta < 360$.

- The value of $\cos(\theta°)$ is the x-coordinate of the intersection point of the terminal ray and the unit circle.
- The value of $\sin(\theta°)$ is the y-coordinate of the intersection point of the terminal ray and the unit circle.
- The sine and cosine functions have domain of all real numbers and range $[-1,1]$.

Problem Set

1. Fill in the chart. Write in the reference angles and the values of the sine and cosine functions for the indicated values of θ.

Amount of rotation, θ, in degrees	Measure of Reference Angle, in degrees	$\cos(\theta°)$	$\sin(\theta°)$
120			
135			
150			
225			
240			
300			
330			

2. Using geometry, Jennifer correctly calculated that $\sin(15°) = \frac{1}{2}\sqrt{2-\sqrt{3}}$. Based on this information, fill in the chart.

Amount of rotation, θ, in degrees	Measure of Reference Angle, in degrees	$\cos(\theta°)$	$\sin(\theta°)$
15			
165			
195			
345			

3. Suppose $0 < \theta < 90$ and $\sin(\theta°) = \frac{1}{\sqrt{3}}$. What is the value of $\cos(\theta°)$?

4. Suppose $90 < \theta < 180$ and $\sin(\theta°) = \frac{1}{\sqrt{3}}$. What is the value of $\cos(\theta°)$?

5. If $\cos(\theta°) = -\frac{1}{\sqrt{5}}$, what are two possible values of $\sin(\theta°)$?

6. Johnny rotated the initial ray through θ degrees, found the intersection of the terminal ray with the unit circle, and calculated that $\sin(\theta°) = \sqrt{2}$. Ernesto insists that Johnny made a mistake in his calculation. Explain why Ernesto is correct.

7. If $\sin(\theta°) = 0.5$, and we know that $\cos(\theta°) < 0$, then what is the smallest possible positive value of θ?

8. The vertices of triangle $\triangle ABC$ have coordinates $A(0,0)$, $B(12,5)$, and $C(12,0)$.
   a. Argue that $\triangle ABC$ is a right triangle.
   b. What are the coordinates where the hypotenuse of $\triangle ABC$ intersects the unit circle $x^2 + y^2 = 1$?
   c. Let θ denote the number of degrees of rotation from $\overrightarrow{AC}$ to $\overrightarrow{AB}$. Calculate $\sin(\theta°)$ and $\cos(\theta°)$.

9. The vertices of triangle $\triangle ABC$ have coordinates $A(0,0)$, $B(4,3)$, and $C(4,0)$. The vertices of triangle $\triangle ADE$ are $A(0,0)$, $D(3,4)$, and $E(3,0)$.
   a. Argue that $\triangle ABC$ is a right triangle.
   b. What are the coordinates where the hypotenuse of $\triangle ABC$ intersects the unit circle $x^2 + y^2 = 1$?
   c. Let θ denote the number of degrees of rotation from $\overrightarrow{AC}$ to $\overrightarrow{AB}$. Calculate $\sin(\theta°)$ and $\cos(\theta°)$.
   d. Argue that $\triangle ADE$ is a right triangle.
   e. What are the coordinates where the hypotenuse of $\triangle ADE$ intersects the unit circle $x^2 + y^2 = 1$?
   f. Let ϕ denote the number of degrees of rotation from $\overrightarrow{AE}$ to $\overrightarrow{AD}$. Calculate $\sin(\phi°)$ and $\cos(\phi°)$.
   g. What is the relation between the sine and cosine of θ and the sine and cosine of ϕ?

10. Use a diagram to explain why $\sin(135°) = \sin(45°)$, but $\cos(135°) \neq \cos(45°)$.

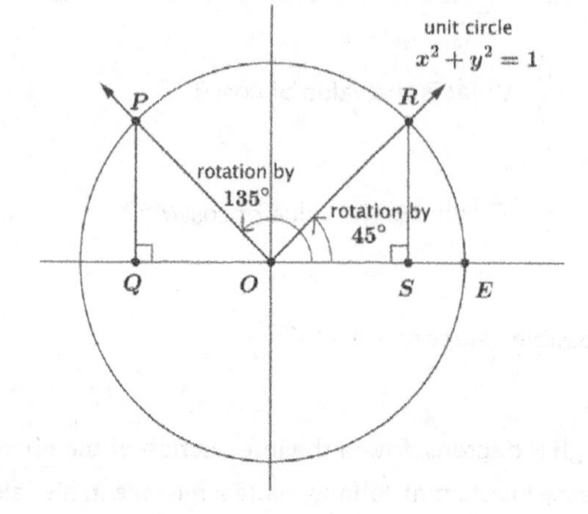

Lesson 5: Extending the Domain of Sine and Cosine to All Real Numbers

Classwork

Opening Exercise

a. Suppose that a group of 360 coworkers pool their money, buying a single lottery ticket every day with the understanding that if any ticket is a winning ticket, the group will split the winnings evenly, and they will donate any leftover money to the local high school. Using this strategy, if the group wins $\$1{,}000$, how much money will be donated to the school?

b. What if the winning ticket is worth $\$250{,}000$? Using the same plan as in part (a), how much money will be donated to the school?

c. What if the winning ticket is worth $\$540{,}000$? Using the same plan as in part (a), how much money will be donated to the school?

Exercises 1–5

1. Find $\cos(405°)$ and $\sin(405°)$. Identify the measure of the reference angle.

2. Find $\cos(840°)$ and $\sin(840°)$. Identify the measure of the reference angle.

3. Find $\cos(1680°)$ and $\sin(1680°)$. Identify the measure of the reference angle.

4. Find $\cos(2115°)$ and $\sin(2115°)$. Identify the measure of the reference angle.

5. Find $\cos(720\,030°)$ and $\sin(720\,030°)$. Identify the measure of the reference angle.

Exercises 6–10

6. Find $\cos(-30°)$ and $\sin(-30°)$. Identify the measure of the reference angle.

7. Find $\cos(-135°)$ and $\sin(-135°)$. Identify the measure of the reference angle.

8. Find $\cos(-1320°)$ and $\sin(-1320°)$. Identify the measure of the reference angle.

9. Find $\cos(-2205°)$ and $\sin(-2205°)$. Identify the measure of the reference angle.

10. Find $\cos(-2835°)$ and $\sin(-2835°)$. Identify the measure of the reference angle.

Discussion

Case 1: What about the values of the sine and cosine function of other amounts of rotation that produce a terminal ray along the positive x-axis, such as $1080°$?

Our definition of a reference angle is the angle formed by the terminal ray and the x-axis, but our terminal ray lies along the x-axis so the terminal ray and the x-axis form a zero angle.

How would we assign values to $\cos(1080°)$ and $\sin(1080°)$?

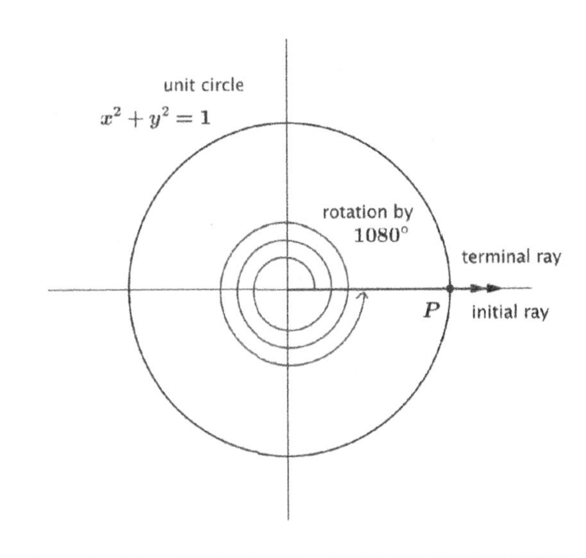

What if we rotated around $24,000°$, which is 400 turns? What are $\cos(24000°)$ and $\sin(24000°)$?

State a generalization of these results:

If $\theta = n \cdot 360$, for some integer n, then $\cos(\theta°) =$ ____, and $\sin(\theta°) =$ ____.

Case 2: What about the values of the sine and cosine function of other amounts of rotation that produce a terminal ray along the negative x-axis, such as $540°$?

How would we assign values to $\cos(540°)$ and $\sin(540°)$?

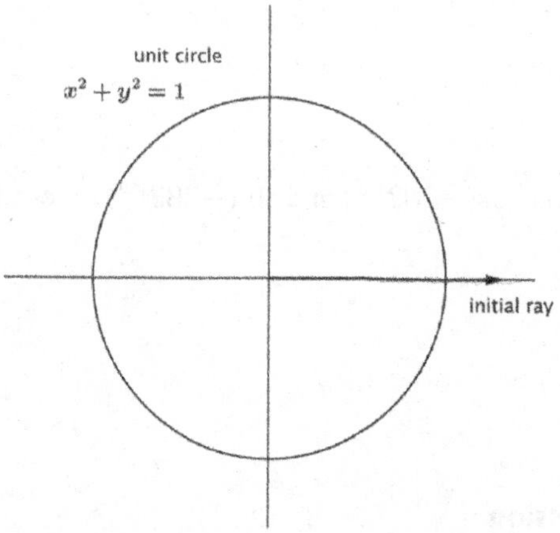

unit circle

$x^2 + y^2 = 1$

initial ray

What are the values of $\cos(900°)$ and $\sin(900°)$? How do you know?

State a generalization of these results:

If $\theta = n \cdot 360 + 180$, for some integer n, then $\cos(\theta°) =$ ____, and $\sin(\theta°) =$ ____.

Case 3: What about the values of the sine and cosine function for rotations that are $90°$ more than a number of full turns, such as $-630°$? How would we assign values to $\cos(-630°)$, and $\sin(-630°)$?

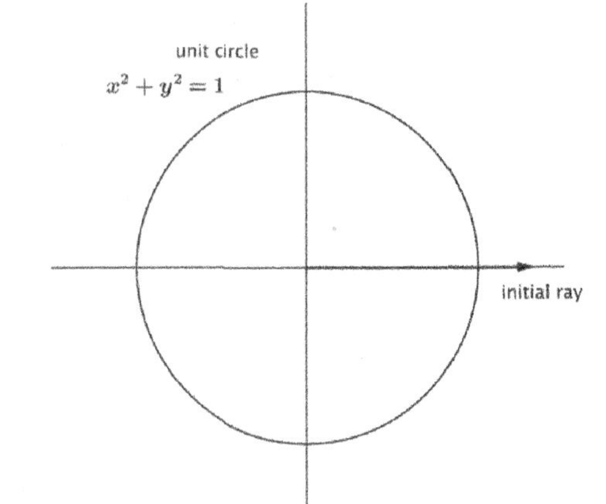

Can we generalize to any rotation that produces a terminal ray along the positive y-axis?

State a generalization of these results:

If $\theta = n \cdot 360 + 90$, *for some integer* n, *then* $\cos(\theta°) =$ ____, *and* $\sin(\theta°) =$ ____.

Case 4: What about the values of the sine and cosine function for rotations whose terminal ray lies along the negative y-axis, such as $-810°$?

How would we assign values to $\cos(-810°)$ and $\sin(-810°)$?

unit circle
$x^2 + y^2 = 1$
rotation by $-810°$
initial ray
P
terminal ray

Can we generalize to any rotation that produces a terminal ray along the negative y-axis?

State a generalization of these results:

If $\theta = n \cdot 360 + 270$, *for some integer* n, *then* $\cos(\theta°) =$ ____, *and* $\sin(\theta°) =$ ____.

Discussion

Let θ be any real number. In the Cartesian plane, rotate the initial ray by θ degrees about the origin. Intersect the resulting terminal ray with the unit circle to get a point (x_θ, y_θ) in the coordinate plane. The value of $\sin(\theta°)$ is y_θ, and the value of $\cos(\theta°)$ is x_θ.

Lesson Summary

In this lesson the definition of the sine and cosine are formalized as functions of a number of degrees of rotation, θ. The initial ray made from the positive x-axis through θ degrees is rotated, going counterclockwise if $\theta > 0$ and clockwise if $\theta < 0$. The point P is defined by the intersection of the terminal ray and the unit circle.

- The value of $\cos(\theta°)$ is the x-coordinate of P.
- The value of $\sin(\theta°)$ is the y-coordinate of P.
- The sine and cosine functions have domain of all real numbers and range $[-1,1]$.

Problem Set

1. Fill in the chart. Write in the measures of the reference angles and the values of the sine and cosine functions for the indicated values of θ.

Number of degrees of rotation, θ	Quadrant	Measure of Reference Angle, in degrees	$\cos(\theta°)$	$\sin(\theta°)$
690				
810				
1560				
1440				
855				
-330				
-4500				
-510				
-135				
-1170				

2. Using geometry, Jennifer correctly calculated that $\sin(15°) = \frac{1}{2}\sqrt{2-\sqrt{3}}$. Based on this information, fill in the chart:

Number of degrees of rotation, θ	Quadrant	Measure of Reference Angle, in degrees	$\cos(\theta°)$	$\sin(\theta°)$
525				
705				
915				
-15				
-165				
-705				

3. Suppose θ represents a number of degrees of rotation and that $\sin(\theta°) = 0.5$. List the first six possible positive values that θ can take.

4. Suppose θ represents a number of degrees of rotation and that $\sin(\theta°) = -0.5$. List six possible negative values that θ can take.

5. Suppose θ represents a number of degrees of rotation. Is it possible that $\cos(\theta°) = \frac{1}{2}$ and $\sin(\theta°) = \frac{1}{2}$?

6. Jane says that since the reference angle for a rotation through $-765°$ has measure $45°$, then $\cos(-765°) = \cos(45°)$, and $\sin(-765°) = \sin(45°)$. Explain why she is or is not correct.

7. Doug says that since the reference angle for a rotation through $765°$ has measure $45°$, then $\cos(765°) = \cos(45°)$, and $\sin(765°) = \sin(45°)$. Explain why he is or is not correct.

Lesson 6: Why Call It Tangent?

Classwork

Opening Exercise

Let $P(x_\theta, y_\theta)$ be the point where the terminal ray intersects the unit circle after rotation by θ degrees, as shown in the diagram below.

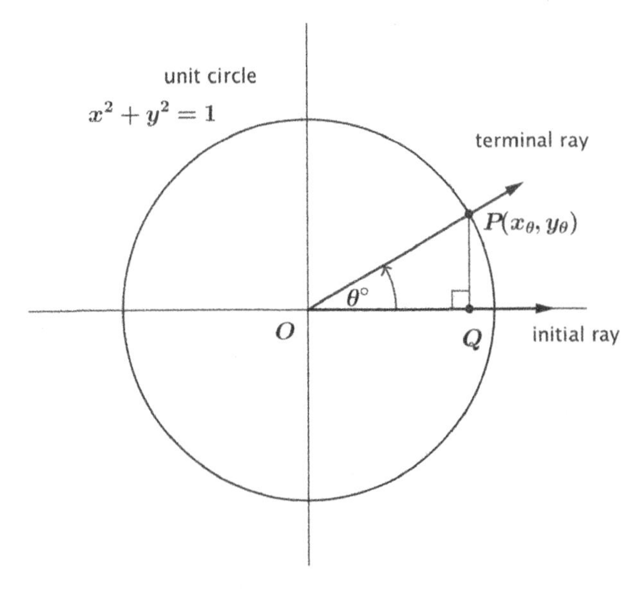

a. Using triangle trigonometry, what are the values of x_θ and y_θ in terms of θ?

b. Using triangle trigonometry, what is the value of $\tan(\theta°)$ in terms of x_θ and y_θ?

c. What is the value of $\tan(\theta°)$ in terms of θ?

Discussion

A description of the tangent function is provided below. Be prepared to answer questions based on your understanding of this function and to discuss your responses with others in your class.

Let θ be any real number. In the Cartesian plane, rotate the nonnegative x-axis by θ degrees about the origin. Intersect the resulting terminal ray with the unit circle to get a point (x_θ, y_θ). If $x_\theta \neq 0$, then the value of $\tan(\theta°)$ is $\frac{y_\theta}{x_\theta}$. In terms of the sine and cosine functions, $\tan(\theta°) = \frac{\sin(\theta°)}{\cos(\theta°)}$ for $\cos(\theta°) \neq 0$.

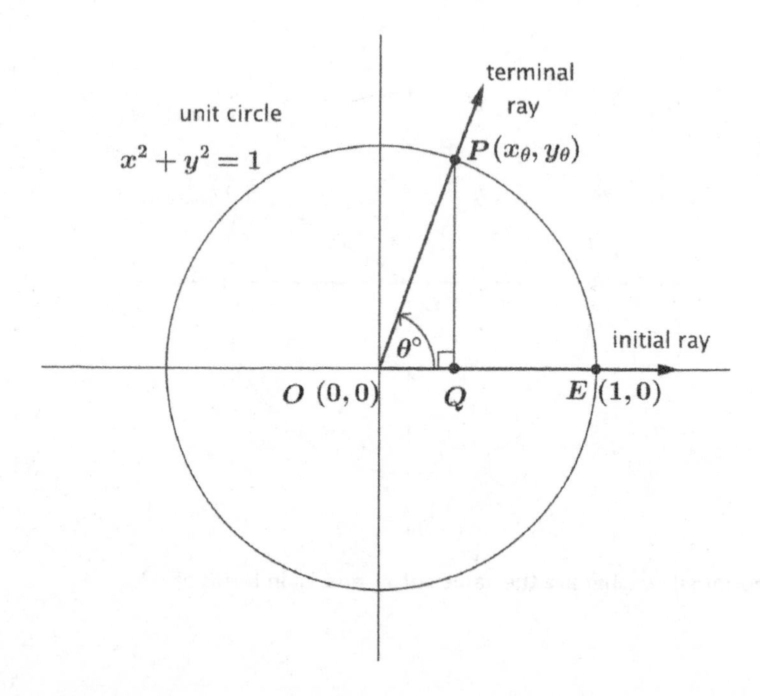

Exercise 1

1. For each value of θ in the table below, use the given values of $\sin(\theta°)$ and $\cos(\theta°)$ to approximate $\tan(\theta°)$ to two decimal places.

θ (degrees)	$\sin(\theta°)$	$\cos(\theta°)$	$\tan(\theta°)$
-89.9	-0.999998	0.00175	
-89	-0.9998	0.0175	
-85	-0.996	0.087	
-80	-0.98	0.17	
-60	-0.87	0.50	
-40	-0.64	0.77	
-20	-0.34	0.94	
0	0	1.00	
20	0.34	0.94	
40	0.64	0.77	
60	0.87	0.50	
80	0.98	0.17	
85	0.996	0.087	
89	0.9998	0.0175	
89.9	0.999998	0.00175	

a. As $\theta \rightarrow -90°$ and $\theta > -90°$, what value does $\sin(\theta°)$ approach?

b. As $\theta \rightarrow -90°$ and $\theta > -90°$, what value does $\cos(\theta°)$ approach?

c. As $\theta \rightarrow -90°$ and $\theta > -90°$, how would you describe the value of $\tan(\theta°) = \frac{\sin(\theta°)}{\cos(\theta°)}$?

d. As $\theta \rightarrow 90°$ and $\theta < 90°$, what value does $\sin(\theta°)$ approach?

e. As $\theta \rightarrow 90°$ and $\theta < 90°$, what value does $\cos(\theta°)$ approach?

f. As $\theta \rightarrow 90°$ and $\theta < 90°$, how would you describe the behavior of $\tan(\theta°) = \frac{\sin(\theta°)}{\cos(\theta°)}$?

g. How can we describe the range of the tangent function?

Example 1

Suppose that point P is the point on the unit circle obtained by rotating the initial ray through $30°$. Find $\tan(30°)$.

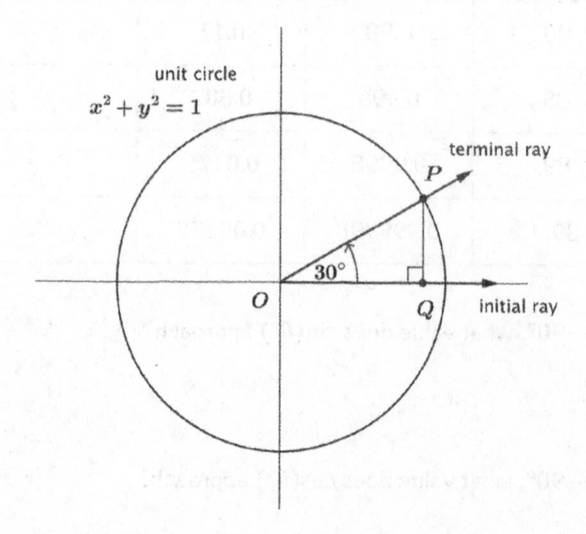

Exercises 2–6: Why Do We Call it Tangent?

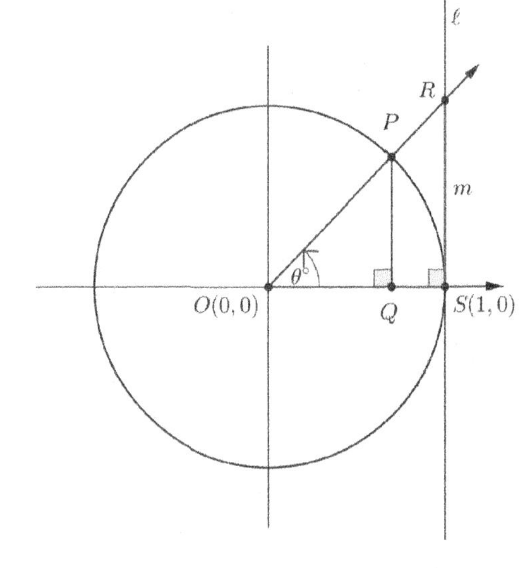

2. Let P be the point on the unit circle with center O that is the intersection of the terminal ray after rotation by θ degrees as shown in the diagram. Let Q be the foot of the perpendicular line from P to the x-axis, and let the line ℓ be the line perpendicular to the x-axis at $S(1,0)$. Let R be the point where the secant line OP intersects the line ℓ. Let m be the length of $\overline{RS}$.

   a. Show that $m = \tan(\theta°)$.

   b. Using a segment in the figure, make a conjecture why mathematicians named the function $f(\theta°) = \frac{\sin(\theta°)}{\cos(\theta°)}$ the tangent function.

   c. Why can you use either triangle, $\triangle POQ$ or $\triangle ROS$, to calculate the length m?

   d. Imagine that you are the mathematician who gets to name the function. (How cool would that be?) Based upon what you know about the equations of lines, what might you have named the function instead?

3. Draw four pictures similar to the diagram in Exercise 2 to illustrate what happens to the value of $\tan(\theta°)$ as the rotation of the secant line through the terminal ray increases towards $90°$. How does your diagram relate to the work done in Exercise 1?

4. When the terminal ray is vertical, what is the relationship between the secant line OR and the tangent line RS? Explain why you cannot determine the measure of m in this instance. What is the value of $\tan(90°)$?

5. When the terminal ray is horizontal, what is the relationship between the secant line OR and the x-axis? Explain what happens to the value of m in this instance. What is the value of $\tan(0°)$?

6. When the terminal ray is rotated counterclockwise about the origin by $45°$, what is the relationship between the value of m and the length of $\overline{OS}$? What is the value of $\tan(45°)$?

Exercises 7–8

7. Rotate the initial ray about the origin the stated number of degrees. Draw a sketch and label the coordinates of point P where the terminal ray intersects the unit circle. What is the slope of the line containing this ray?

   a. $30°$

   b. $45°$

   c. $60°$

d. Use the definition of tangent to find $\tan(30°)$, $\tan(45°)$, and $\tan(60°)$. How do your answers compare your work in parts (a)–(c)?

e. If the initial ray is rotated θ degrees about the origin, show that the slope of the line containing the terminal ray is equal to $\tan(\theta°)$. Explain your reasoning.

f. Now that you have shown that the value of the tangent function is equal to the slope of the terminal ray, would you prefer using the name *tangent function* or *slope function*? Why do you think we use ***tangent*** instead of *slope* as the name of the tangent function?

8. Rotate the initial ray about the origin the stated number of degrees. Draw a sketch and label the coordinates of point P where the terminal ray intersects the unit circle. How does your diagram in this exercise relate to the diagram in the corresponding part of Exercise 7? What is $\tan(\theta°)$ for these values of θ?

a. $210°$

b. $225°$

c. $240°$

d. What do the results of parts (a)–(c) suggest about the value of the tangent function after rotating an additional 180 degrees?

e. What is the period of the tangent function? Discuss with a classmate and write your conclusions.

f. Use the results of Exercise 7(e) to explain why $\tan(0°) = 0$.

g. Use the results of Exercise 7(e) to explain why $\tan(90°)$ is undefined.

Lesson Summary

- A working definition of the tangent function is $\tan(\theta°) = \frac{\sin(\theta°)}{\cos(\theta°)}$, where $\cos(\theta°) \neq 0$.
- The value of $\tan(\theta°)$ is the length of the line segment on the tangent line to the unit circle centered at the origin from the intersection with the unit circle and the intersection with the secant line created by the x-axis rotated θ degrees. (This is why we call it tangent.)
- The value of $\tan(\theta°)$ is the slope of the line obtained by rotating the x-axis θ degrees about the origin.
- The domain of the tangent function is $\{\theta \in \mathbb{R} | \theta \neq 90 + 180k, \text{for all integers } k\}$ which is equivalent to $\{\theta \in \mathbb{R} | \cos(\theta°) \neq 0\}$.
- The range of the tangent function is all real numbers.
- The period of the tangent function is $180°$.

$\tan(0°)$	$\tan(30°)$	$\tan(45°)$	$\tan(60°)$	$\tan(90°)$
0	$\frac{\sqrt{3}}{3}$	1	$\sqrt{3}$	undefined

Problem Set

1. Label the missing side lengths, and find the value of $\tan(\theta°)$ in the following right triangles.

   a. $\theta = 30$

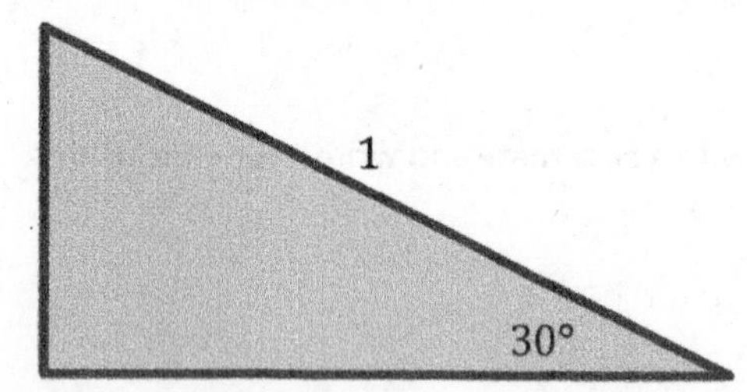

   b. $\theta = 45$

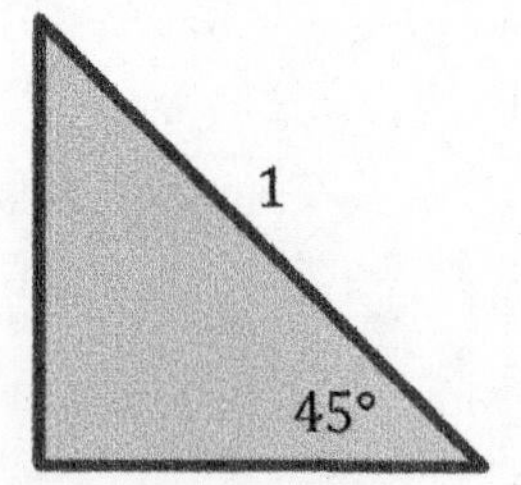

c. $\theta = 60$

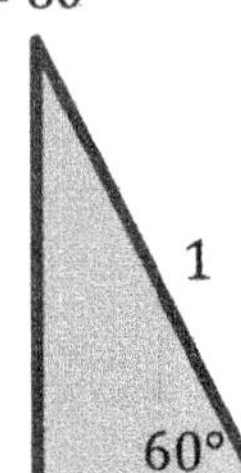

2. Let θ be any real number. In the Cartesian plane, rotate the initial ray by θ degrees about the origin. Intersect the resulting terminal ray with the unit circle to get point $P(x_\theta, y_\theta)$.

a. Complete the table by finding the slope of the line through the origin and the point P.

θ, in degrees	Slope	θ, in degrees	Slope
0		180	
30		210	
45		225	
60		240	
90		270	
120		300	
135		315	
150		330	

b. Explain how these slopes are related to the tangent function.

3. Consider the following diagram of a circle of radius r centered at the origin. The line ℓ is tangent to the circle at $S(r, 0)$, so ℓ is perpendicular to the x-axis.

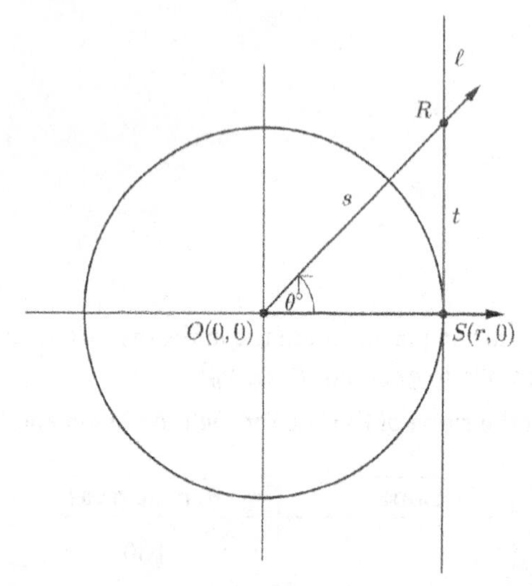

a. If $r = 1$, then state the value of t in terms of one of the trigonometric functions.

b. If r is any positive value, then state the value of t in terms of one of the trigonometric functions.

For the given values of r and θ, find t.

c. $\theta = 30, r = 2$

d. $\theta = 45, r = 2$

e. $\theta = 60, r = 2$

f. $\theta = 45, r = 4$

g. $\theta = 30, r = 3.5$

h. $\theta = 0, r = 9$

i. $\theta = 90, r = 5$

j. $\theta = 60, r = \sqrt{3}$

k. $\theta = 30, r = 2.1$

l. $\theta = A, r = 3$

m. $\theta = 30, r = b$

n. Knowing that $\tan(\theta°) = \frac{\sin(\theta°)}{\cos(\theta°)}$, for $r = 1$, find the value of s in terms of one of the trigonometric functions.

4. Using what you know of the tangent function, show that $-\tan(\theta°) = \tan(-\theta°)$ for $\theta \neq 90 + 180k$, for all integers k.

Lesson 7: Secant and the Co-Functions

Classwork

Opening Exercise

Find the length of each segment below in terms of the value of a trigonometric function.

$OQ =$ __________ $PQ =$ __________ $RS =$ __________

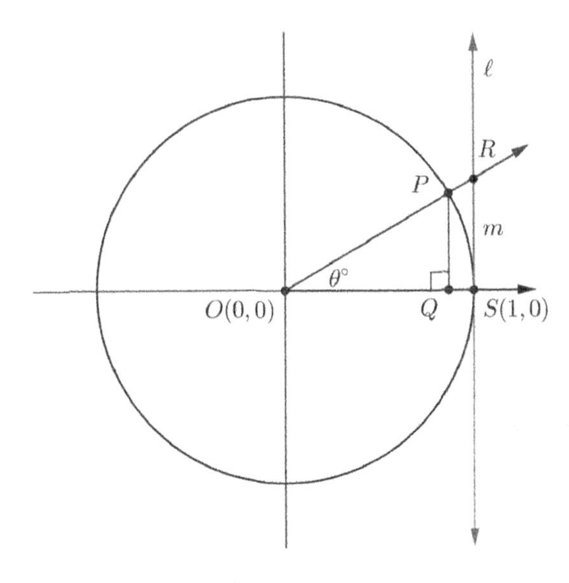

Example 1

Use similar triangles to find the value of $\sec(\theta°)$ in terms of one other trigonometric function.

Exercise 1

The definition of the secant function is offered below. Answer the questions to better understand this definition and the domain and range of this function. Be prepared to discuss your responses with others in your class.

> Let θ be any real number.
> In the Cartesian plane, rotate the nonnegative x-axis by θ degrees about the origin. Intersect this new ray with the unit circle to get a point (x_θ, y_θ).
>
> If $x_\theta \neq 0$, then the value of $\sec(\theta°)$ is $\frac{1}{x_\theta}$.
>
> Otherwise, $\sec(\theta°)$ is undefined.
>
> In terms of the cosine function, $\sec(\theta°) = \frac{1}{\cos(\theta°)}$ for $\cos(\theta°) \neq 0$.

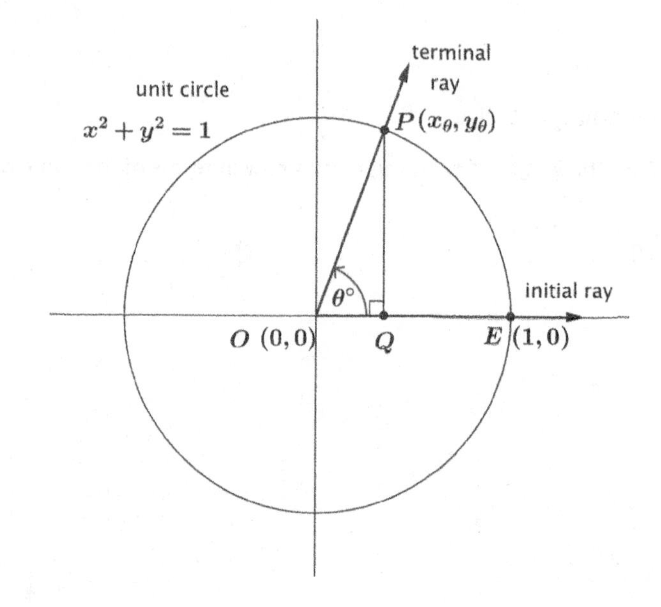

a. What is the domain of the secant function?

b. The domains of the secant and tangent functions are the same. Why?

c. What is the range of the secant function? How is this range related to the range of the cosine function?

d. Is the secant function a periodic function? If so, what is its period?

Exercise 2

In the diagram, the horizontal blue line is tangent to the unit circle at $(0,1)$.

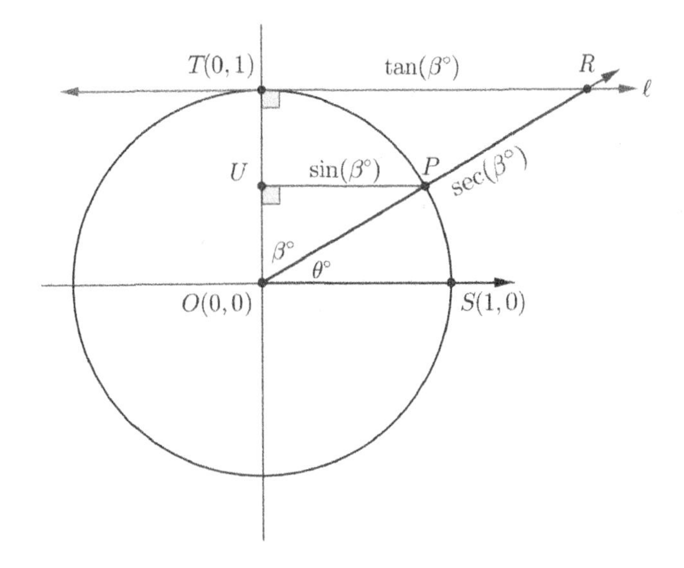

a. How does this diagram compare to the one given in the Opening Exercise?

b. What is the relationship between β and θ?

c. Which segment in the figure has length $\sin(\theta°)$? Which segment has length $\cos(\theta°)$?

d. Which segment in the figure has length $\sin(\beta°)$? Which segment has length $\cos(\beta°)$?

e. How can you write $\sin(\theta°)$ and $\cos(\theta°)$ in terms of the trigonometric functions of β?

Example 2

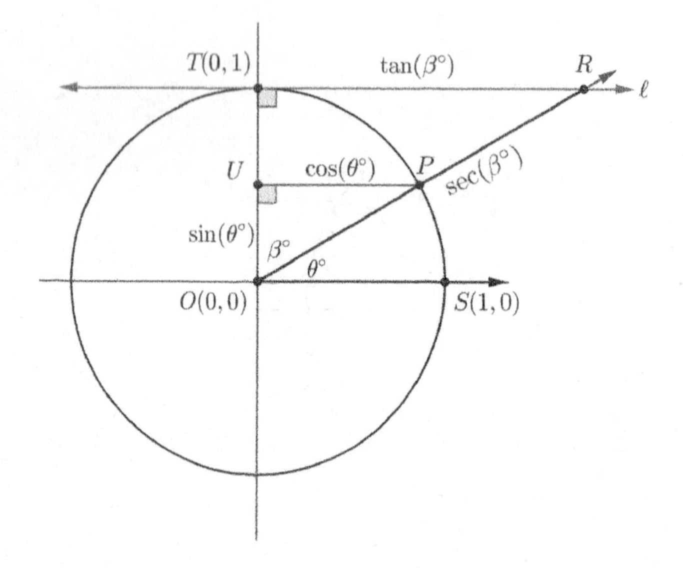

The horizontal blue line is tangent to the circle at $(0,1)$.

a. If two angles are complements with measures β and θ as shown in the diagram, use similar triangles to show that $\sec(\beta°) = \frac{1}{\sin(\theta°)}$.

b. If two angles are complements with measures β and θ as shown in the diagram, use similar triangles to show that $\tan(\beta°) = \frac{1}{\tan(\theta°)}$.

Discussion

Definitions of the cosecant and cotangent functions are offered below. Answer the questions to better understand the definitions and the domains and ranges of these functions. Be prepared to discuss your responses with others in your class.

> Let θ be any real number such that $\theta \neq 180k$ for all integers k.
>
> In the Cartesian plane, rotate the initial ray by θ degrees about the origin. Intersect the resulting terminal ray with the unit circle to get a point (x_θ, y_θ).
>
> The value of $\csc(\theta°)$ is $\frac{1}{y_\theta}$.
>
> The value of $\cot(\theta°)$ is $\frac{x_\theta}{y_\theta}$.

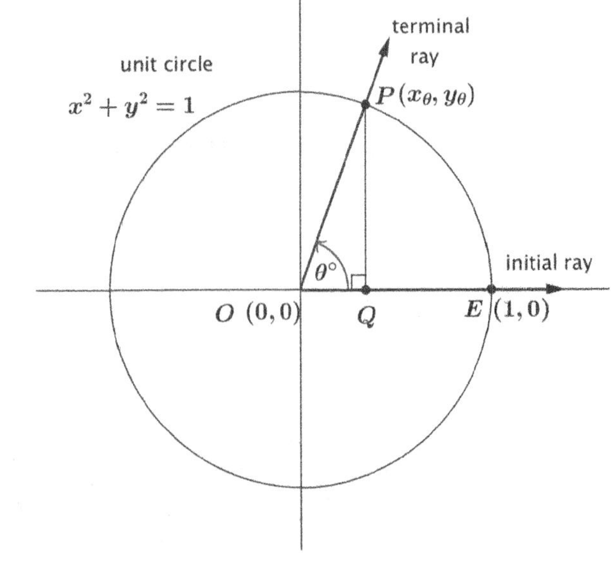

The secant, cosecant, and cotangent functions are often referred to as reciprocal functions. Why do you think these functions are so named?

Why are the domains of these functions restricted?

The domains of the cosecant and cotangent functions are the same. Why?

What is the range of the cosecant function? How is this range related to the range of the sine function?

What is the range of the cotangent function? How is this range related to the range of the tangent function?

Let θ be any real number. In the Cartesian plane, rotate the initial ray by θ degrees about the origin. Intersect the resulting terminal ray with the unit circle to get a point (x_θ, y_θ). Then:			
Function	**Value**	**For any θ such that...**	**Formula**
Sine	y_θ	θ is a real number	
Cosine	x_θ	θ is a real number	
Tangent	$\frac{y_\theta}{x_\theta}$	$\theta \neq 90 + 180k$, for all integers k	$\tan(\theta°) = \frac{\sin(\theta°)}{\cos(\theta°)}$
Secant	$\frac{1}{x_\theta}$	$\theta \neq 90 + 180k$, for all integers k	$\sec(\theta°) = \frac{1}{\cos(\theta°)}$
Cosecant	$\frac{1}{y_\theta}$	$\theta \neq 180k$, for all integers k	$\csc(\theta°) = \frac{1}{\sin(\theta°)}$
Cotangent	$\frac{x_\theta}{y_\theta}$	$\theta \neq 180k$, for all integers k	$\cot(\theta°) = \frac{\cos(\theta°)}{\sin(\theta°)}$

Problem Set

1. Use the reciprocal interpretations of $\sec(\theta°)$, $\csc(\theta°)$, and $\cot(\theta°)$ and the unit circle provided to complete the table.

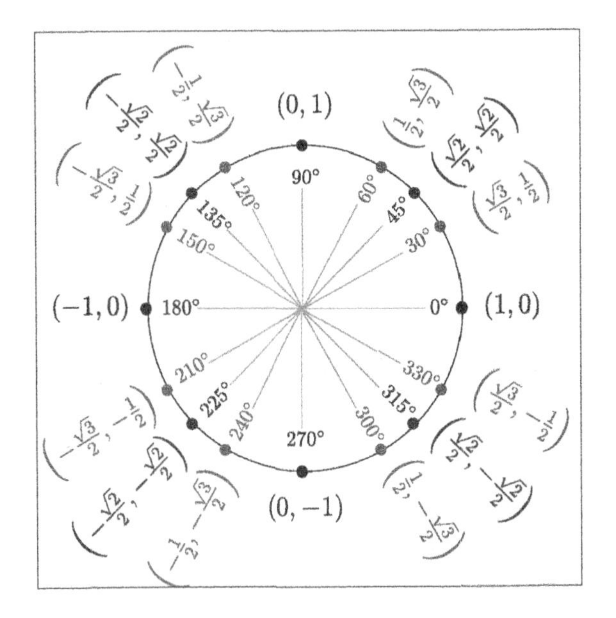

θ, in degrees	$\sec(\theta°)$	$\csc(\theta°)$	$\cot(\theta°)$
0			
30			
45			
60			
90			
120			
180			
225			
240			
270			
315			
330			

2. Find the following values from the information given.
   a. $\sec(\theta°)$; $\cos(\theta°) = 0.3$
   b. $\csc(\theta°)$; $\sin(\theta°) = -0.05$
   c. $\cot(\theta°)$; $\tan(\theta°) = 1000$
   d. $\sec(\theta°)$; $\cos(\theta°) = -0.9$
   e. $\csc(\theta°)$; $\sin(\theta°) = 0$
   f. $\cot(\theta°)$; $\tan(\theta°) = -0.0005$

3. Choose three θ values from the table in Problem 1 for which $\sec(\theta°)$, $\csc(\theta°)$, and $\tan(\theta°)$ are defined and not zero. Show that for these values of θ, $\frac{\sec(\theta°)}{\csc(\theta°)} = \tan(\theta°)$.

4. Find the value of $\sec(\theta°)\cos(\theta°)$ for the following values of θ.
   a. $\theta = 120$
   b. $\theta = 225$
   c. $\theta = 330$
   d. Explain the reasons for the pattern you see in your responses to parts (a)–(c).

5. Draw a diagram representing the two values of θ between 0 and 360 so that $\sin(\theta°) = -\frac{\sqrt{3}}{2}$. Find the values of $\tan(\theta°)$, $\sec(\theta°)$, and $\csc(\theta°)$ for each value of θ.

6. Find the value of $\left(\sec(\theta°)\right)^2 - \left(\tan(\theta°)\right)^2$ when $\theta = 225$.

7. Find the value of $\left(\csc(\theta°)\right)^2 - \left(\cot(\theta°)\right)^2$ when $\theta = 330$.

Extension:

8. Using the formulas $\sec(\theta°) = \frac{1}{\cos(\theta°)}$, $\csc(\theta°) = \frac{1}{\sin(\theta°)}$, and $\cot(\theta°) = \frac{1}{\tan(\theta°)}$, show that $\frac{\sec(\theta°)}{\csc(\theta°)} = \tan(\theta°)$, where these functions are defined and not zero.

9. Tara showed that $\frac{\sec(\theta°)}{\csc(\theta°)} = \tan(\theta°)$, for values of θ for which the functions are defined and $\csc(\theta°) \neq 0$, and then concluded that $\sec(\theta°) = \sin(\theta°)$ and $\csc(\theta°) = \cos(\theta°)$. Explain what is wrong with her reasoning.

10. From Lesson 6, Ren remembered that the tangent function is odd, meaning that $-\tan(\theta°) = \tan(-\theta°)$ for all θ in the domain of the tangent function. He concluded because of the relationship between the secant function, cosecant function, and tangent function developed in Problem 9, it is impossible for both the secant and the cosecant functions to be odd. Explain why he is correct.

Lesson 8: Graphing the Sine and Cosine Functions

Classwork

Exploratory Challenge 1

Your group will be graphing: $f(\theta) = \sin(\theta°)$ $g(\theta) = \cos(\theta°)$

The circle on the next page is a unit circle, meaning that the length of the radius is one unit.

1. Mark axes on the poster board, with a horizontal axis in the middle of the board and a vertical axis near the left edge, as shown.

2. Measure the radius of the circle using a ruler. Use the length of the radius to mark 1 and -1 on the vertical axis.
3. Wrap the yarn around the circumference of the circle starting at 0. Mark each $15°$ increment on the yarn with the marker. Unwind the yarn and lay it on the horizontal axis. Transfer the marks on the yarn to corresponding increments on the horizontal axis. Label these marks as $0, 15, 30, \ldots, 360$.
4. Record the number of degrees of rotation θ on the horizontal axis of the graph, and record the value of either $\sin(\theta°)$ or $\cos(\theta°)$ on the vertical axis. Notice that the scale is wildly different on the vertical and horizontal axes.
5. If you are graphing $g(\theta) = \cos(\theta°)$: For each θ marked on your horizontal axis, beginning at 0, use the spaghetti to measure the *horizontal* displacement from the vertical axis to the relevant point on the unit circle. The horizontal displacement is the value of the cosine function. Break the spaghetti to mark the correct length, and place it vertically at the appropriate tick mark on the horizontal axis.
6. If you are graphing $f(\theta) = \sin(\theta°)$: For each θ marked on your horizontal axis, beginning at 0, use the spaghetti to measure the *vertical* displacement from the horizontal to the relevant point on the unit circle. The vertical displacement is the value of the sine function. Break the spaghetti to mark the correct length, and place it vertically at the appropriate tick mark on the horizontal axis.
7. Remember to place the spaghetti below the horizontal axis when the value of the sine function or the cosine function is negative. Glue each piece of spaghetti in place.
8. Draw a smooth curve that connects the points at the end of each piece of spaghetti.

Exploratory Challenge 2

Part I: Consider the function $f(\theta) = \sin(\theta°)$.

a. Complete the following table by using the special values learned in Lesson 4. Give values as approximations to one decimal place.

θ, in degrees	0	30	45	60	90	120	135	150	180
$\sin(\theta°)$									

θ, in degrees	210	225	240	270	300	315	330	360
$\sin(\theta°)$								

b. Using the values in the table, sketch the graph of the sine function on the interval $[0, 360]$.

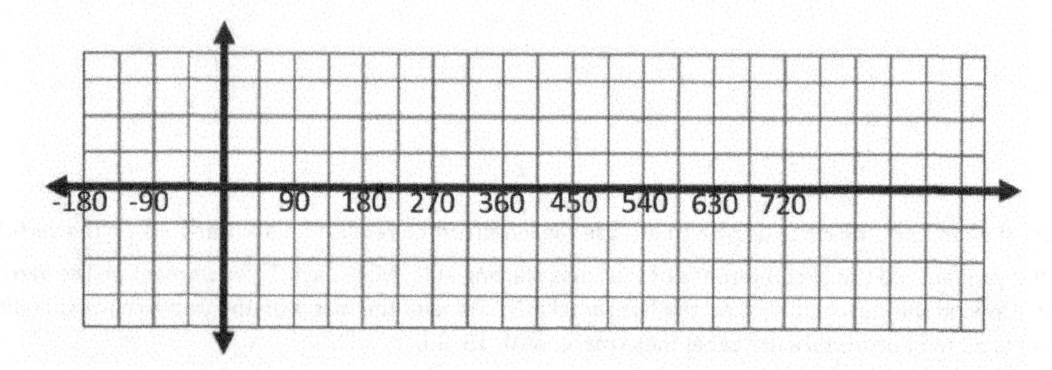

c. Extend the graph of the sine function above so that it is graphed on the interval from $[-180, 720]$.

d. For the interval $[-180, 720]$, describe the values of θ at which the sine function has relative maxima and minima.

e. For the interval $[-180, 720]$, describe the values of θ for which the sine function is increasing and decreasing.

f. For the interval $[-180, 720]$, list the values of θ at which the graph of the sine function crosses the horizontal axis.

g. Describe the end behavior of the sine function.

h. Based on the graph, is sine an odd function, even function, or neither? How do you know?

i. Describe how the sine function repeats.

Part II: Consider the function $g(\theta) = \cos(\theta°)$.

a. Complete the following table giving answers as approximations to one decimal place.

θ, in degrees	0	30	45	60	90	120	135	150	180
$\cos(\theta°)$									

θ, in degrees	210	225	240	270	300	315	330	360
$\cos(\theta°)$								

b. Using the values in the table, sketch the graph of the cosine function on the interval $[0, 360]$.

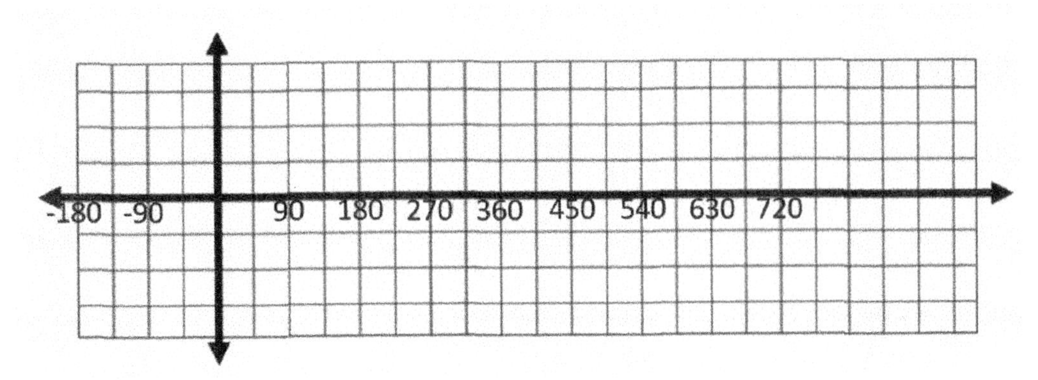

c. Extend the graph of the cosine function above so that it is graphed on the interval from $[-180, 720]$.

d. For the interval $[-180, 270]$, describe the values of θ at which the cosine function has relative maxima and minima.

e. For the interval $[-180, 720]$, describe the values of θ for which the cosine function is increasing and decreasing.

f. For the interval $[-180, 720]$, list the values of θ at which the graph of the cosine function crosses the horizontal axis.

g. Describe the end behavior of the graph of the cosine function.

h. Based on the graph, is cosine an odd function, even function, or neither? How do you know?

i. Describe how the cosine function repeats.

j. How are the sine function and the cosine function related to each other?

Lesson Summary

- A function f whose domain is a subset of the real numbers is said to be *periodic with period* $P > 0$ if the domain of f contains $x + P$ whenever it contains x, and if $f(x + P) = f(x)$ for all real numbers x in its domain.
- If a least positive number P exists that satisfies this equation, it is called the *fundamental period* or, if the context is clear, just the *period* of the function.
- The *amplitude* of the sine or cosine function is half of the distance between a maximal point and a minimal point of the graph of the function.

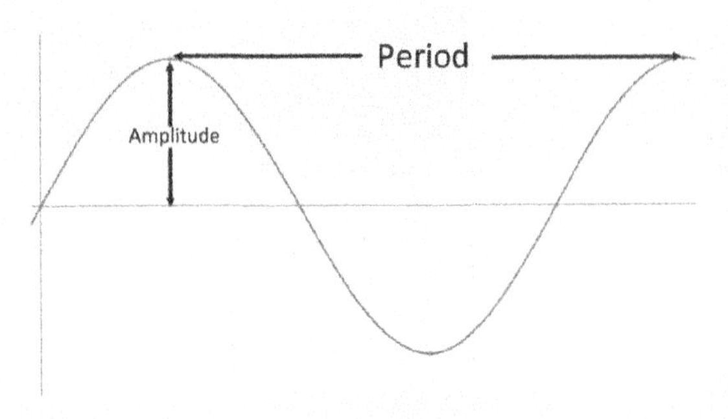

Problem Set

1. Graph the sine function on the interval $[-360, 360]$ showing all key points of the graph (horizontal and vertical intercepts and maximum and minimum points). Then, use the graph to answer each of the following questions.
   a. On the interval $[-360, 360]$, what are the relative minima of the sine function? Why?
   b. On the interval $[-360, 360]$, what are the relative maxima of the sine function? Why?
   c. On the interval $[-360, 360]$, for what values of θ is $\sin(\theta°) = 0$? Why?
   d. If we continued to extend the graph in either direction, what would it look like? Why?
   e. Arrange the following values in order from smallest to largest by using their location on the graph.

      $\sin(170°)$ $\sin(85°)$ $\sin(-85°)$ $\sin(200°)$

   f. On the interval $(90, 270)$, is the graph of the sine function increasing or decreasing? Based on that, name another interval not included in $(90, 270)$ where the sine function must have the same behavior.

2. Graph the cosine function on the interval $[-360, 360]$ showing all key points of the graph (horizontal and vertical intercepts and maximum and minimum points). Then, use the graph to answer each of the following questions.
   a. On the interval $[-360, 360]$, what are the relative minima of the cosine function? Why?
   b. On the interval $[-360, 360]$, what are the relative maxima of the cosine function? Why?
   c. On the interval $[-360, 360]$, for what values of θ is $\cos(\theta°) = 0$? Why?
   d. If we continued to extend the graph in either direction, what would it look like? Why?
   e. What can be said about the end behavior of the cosine function?
   f. Arrange the following values in order from smallest to largest by using their location on the graph.

   $\cos(135°)$ $\cos(85°)$ $\cos(-15°)$ $\cos(190°)$

3. Write a paragraph comparing and contrasting the sine and cosine functions using their graphs and end behavior.

4. Use the graph of the sine function given below to answer the following questions.

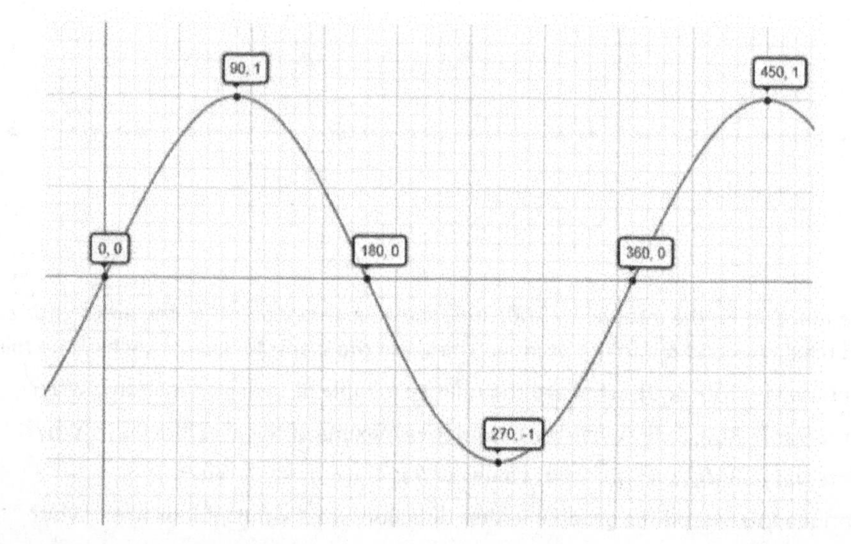

   a. Desmond is trying to determine the value of $\sin(45°)$. He decides that since 45 is halfway between 0 and 90 that $\sin(45°) = \frac{1}{2}$. Use the graph to show him that he is incorrect.
   b. Using the graph, complete each statement by filling in the symbol >, <, or =.
      i. $\sin(250°)$ ☐ $\sin(290°)$
      ii. $\sin(25°)$ ☐ $\sin(85°)$
      iii. $\sin(140°)$ ☐ $\sin(160°)$
   c. On the interval $[0,450]$, list the values of θ such that $\sin(\theta°) = \frac{1}{2}$.
   d. Explain why there are no values of θ such that $\sin(\theta°) = 2$.

Exploratory Challenge Unit Circle Diagram

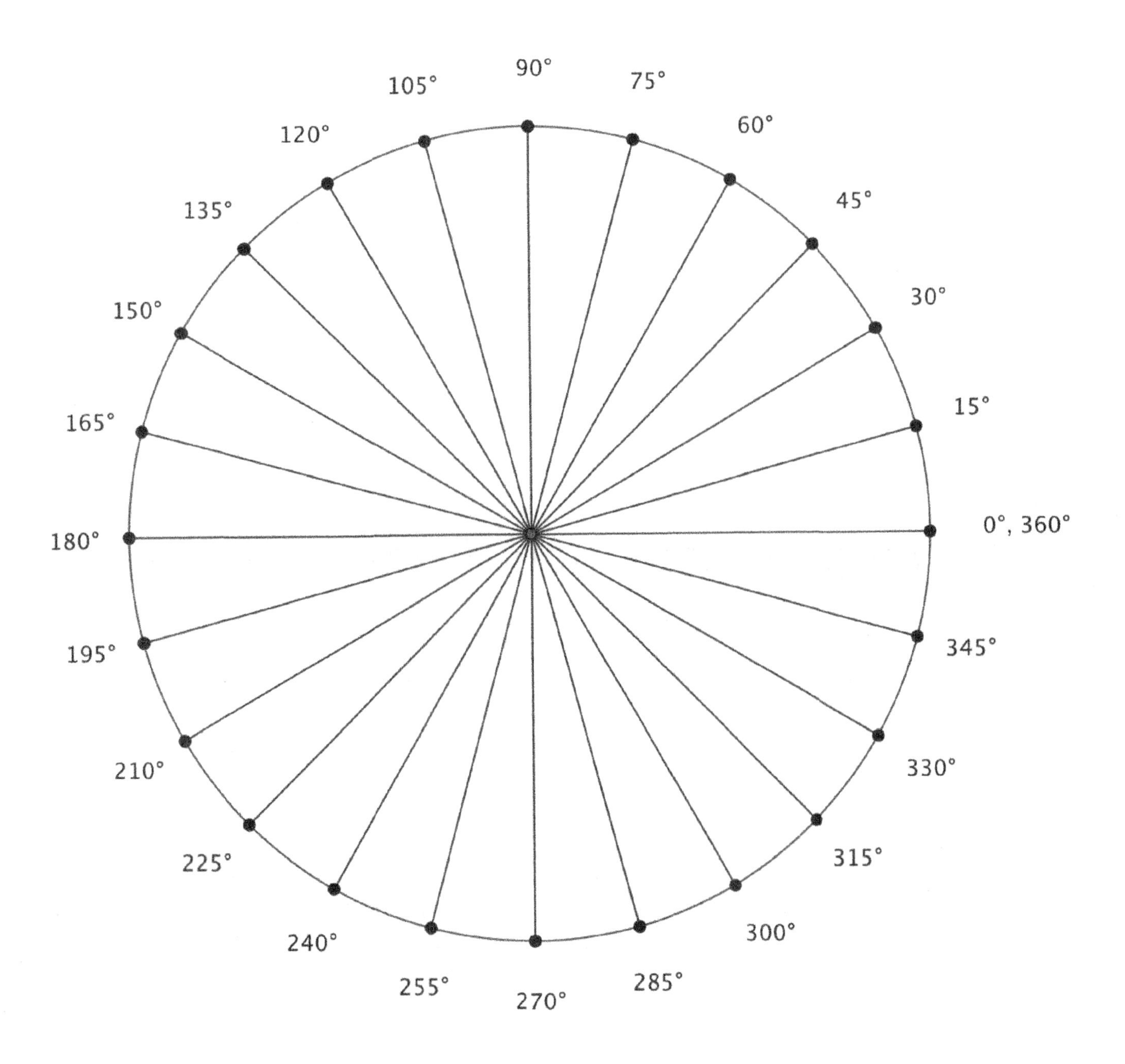

This page intentionally left blank

Lesson 9: Awkward! Who Chose the Number 360, Anyway?

Classwork

Opening Exercise

Let's construct the graph of the function $y = \sin(x°)$, where x is the measure of degrees of rotation. In Lesson 5, we decided that the domain of the sine function is all real numbers and the range is $[-1,1]$. Use your calculator to complete the table below with values rounded to one decimal place, and then graph the function on the axes below. Be sure that your calculator is in degree mode.

x, in degrees	$y = \sin(x°)$
0	
30	
45	
60	
90	
120	

x, in degrees	$y = \sin(x°)$
135	
150	
180	
210	
225	
240	

x, in degrees	$y = \sin(x°)$
270	
300	
315	
330	
360	

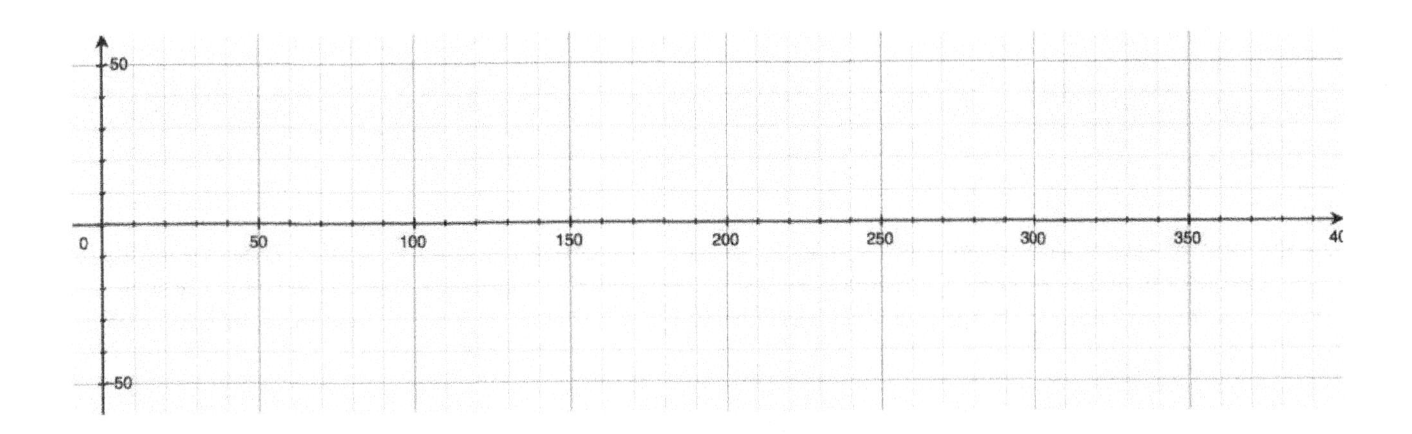

Exercises 1–5

Set your calculator's viewing window to $0 \leq x \leq 10$ and $-2.4 \leq y \leq 2.4$, and be sure that your calculator is in degree mode. Plot the following functions in the same window:

$y = \sin(x°)$

$y = \sin(2x°)$

$y = \sin(10x°)$

$y = \sin(50x°)$

$y = \sin(100x°)$

1. This viewing window was chosen because it has close to the same scale in the horizontal and vertical directions. In this viewing window, which of the five transformed sine functions most clearly shows the behavior of the sine function?

2. Describe the relationship between the steepness of the graph $y = \sin(kx°)$ near the origin and the value of k.

3. Since we can control the steepness of the graph $y = \sin(kx°)$ near the origin by changing the value of k, how steep might we want this graph to be? What is your *favorite* positive slope for a line through the origin?

4. In the same viewing window on your calculator, plot $y = x$ and $y = \sin(kx°)$ for some value of k. Experiment with your calculator to find a value of k so that the steepness of $y = \sin(kx°)$ matches the slope of the line $y = x$ near the origin. You may need to change your viewing window to $0 \leq x \leq 2$ and $0 \leq y \leq 1$ to determine the best value of k.

- A circle is defined by a point and a radius. If we start with a circle of any radius and look at a sector of that circle with an arc length equal to the length of the radius, then the central angle of that sector is always the same size. We define a *radian* to be the measure of that central angle and denote it by 1 rad.

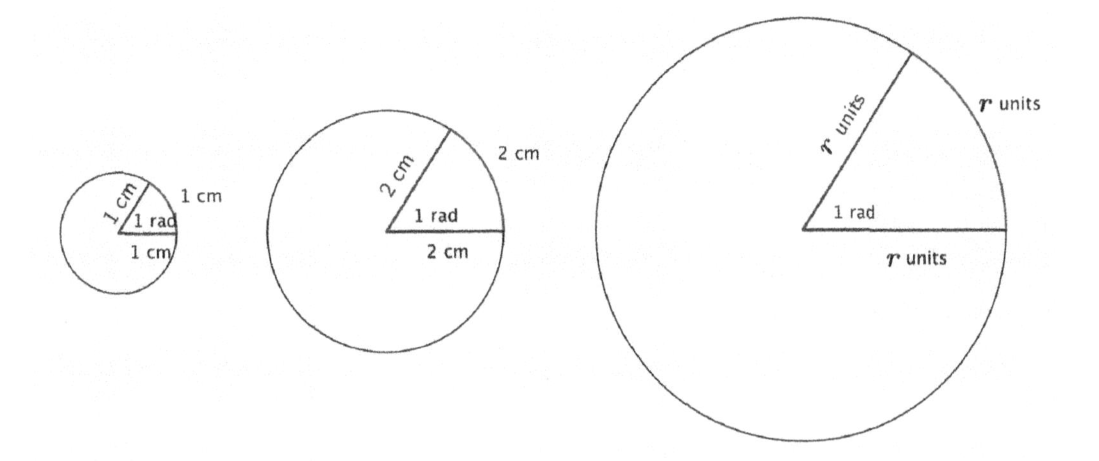

- Thus, a radian measures how far one radius will wrap around the circle. For any circle, it takes $2\pi \approx 6.3$ radius lengths to wrap around the circumference. In the figure, 6 radius lengths are shown around the circle, with roughly 0.3 radius lengths left over.

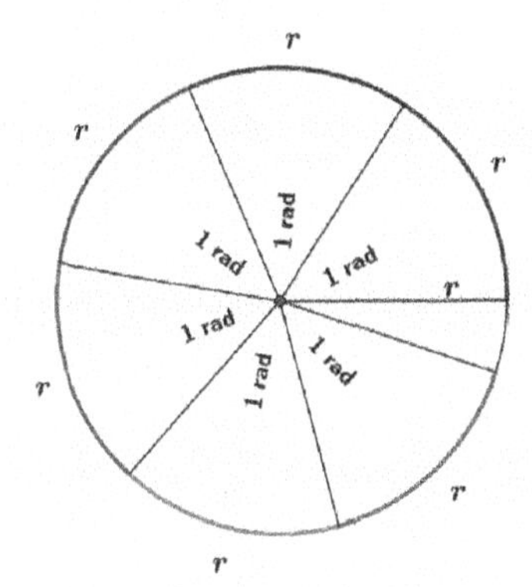

5. Use a protractor that measures angles in degrees to find an approximate degree measure for an angle with measure 1 rad. Use one of the figures from the previous discussion.

Examples 1–4

1. Convert from degrees to radians: $45°$

2. Convert from degrees to radians: $33°$

3. Convert from radians to degrees: $-\frac{\pi}{3}$ rad

4. Convert from radians to degrees: $\frac{19\pi}{17}$ rad

Exercises 6–7

6. Complete the table below, converting from degrees to radians or from radians to degrees as necessary. Leave your answers in exact form, involving π.

Degrees	Radians
$45°$	$\frac{\pi}{4}$
$120°$	
	$-\frac{5\pi}{6}$
	$\frac{3\pi}{2}$
$450°$	
$x°$	
	x

7. On your calculator, graph the functions $y = x$ and $y = \sin\left(\frac{180}{\pi}x°\right)$. What do you notice near the origin? What is the decimal approximation to the constant $\frac{180}{\pi}$ to one decimal place? Explain how this relates to what we've done in Exercise 4.

Discussion

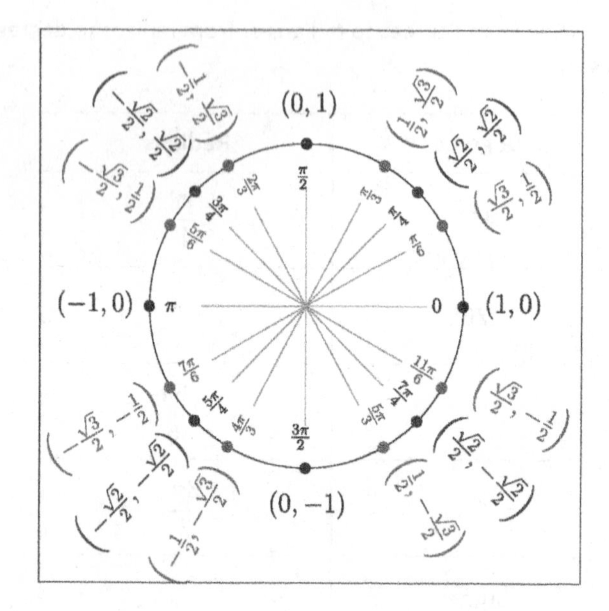

Lesson Summary

- A *radian* is the measure of the central angle of a sector of a circle with arc length of one radius length.
- There are 2π radians in a $360°$ rotation, also known as a *turn*, so degrees are converted to radians and radians to degrees by:

$$2\pi \text{ rad} = 1 \text{ turn} = 360\,°.$$

- **SINE FUNCTION (description)**: The *sine function*, $\sin: \mathbb{R} \to \mathbb{R}$, can be defined as follows: Let θ be any real number. In the Cartesian plane, rotate the initial ray by θ radians about the origin. Intersect the resulting terminal ray with the unit circle to get a point (x_θ, y_θ). The value of $\sin(\theta)$ is y_θ.
- **COSINE FUNCTION (description)**: The *cosine function*, $\cos: \mathbb{R} \to \mathbb{R}$, can be defined as follows: Let θ be any real number. In the Cartesian plane, rotate the initial ray by θ radians about the origin. Intersect the resulting terminal ray with the unit circle to get a point (x_θ, y_θ). The value of $\cos(\theta)$ is x_θ.

Problem Set

1. Use a radian protractor to measure the amount of rotation in radians of ray BA to ray BC in the indicated direction. Measure to the nearest 0.1 radian. Use negative measures to indicate clockwise rotation.

a.

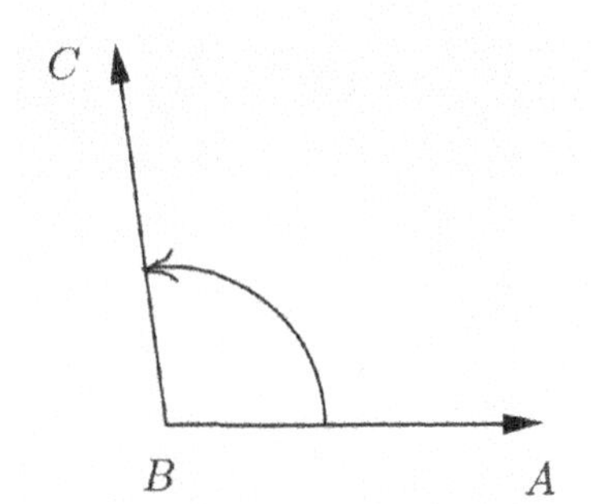

b.

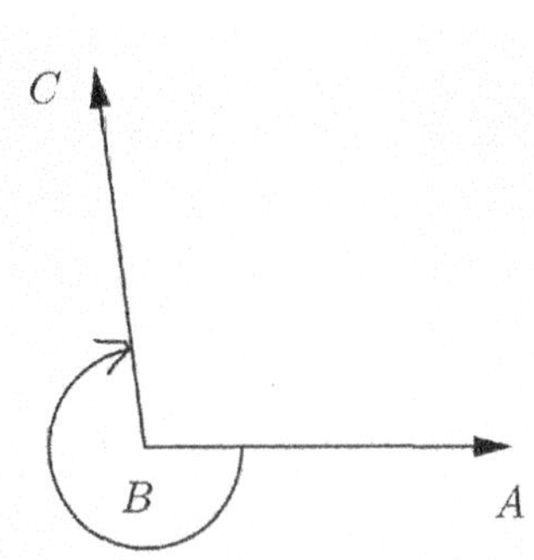

c.

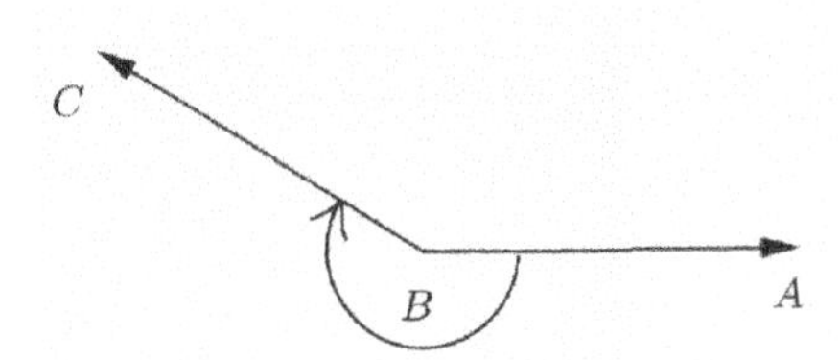

d.

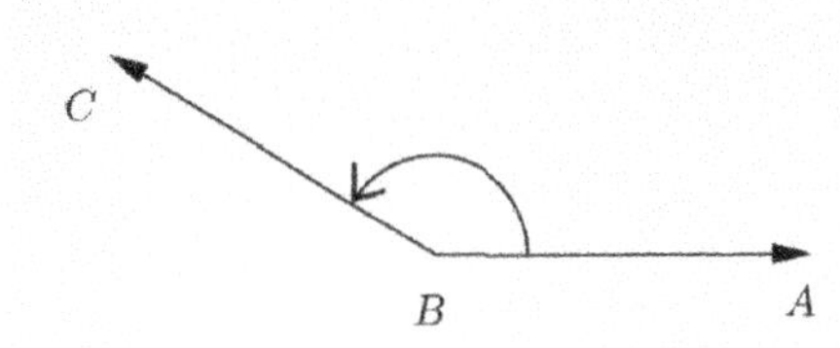

e.

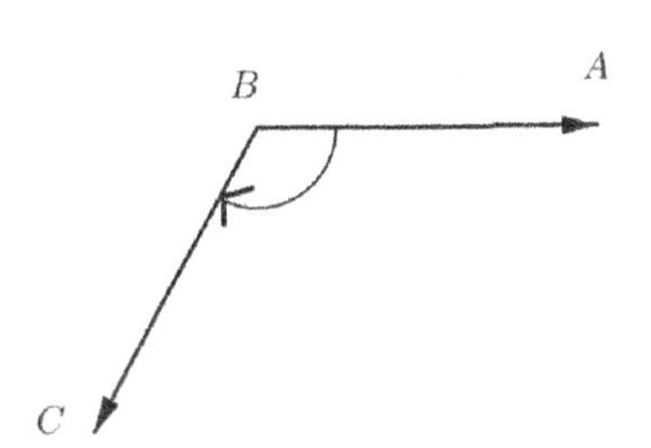

f.

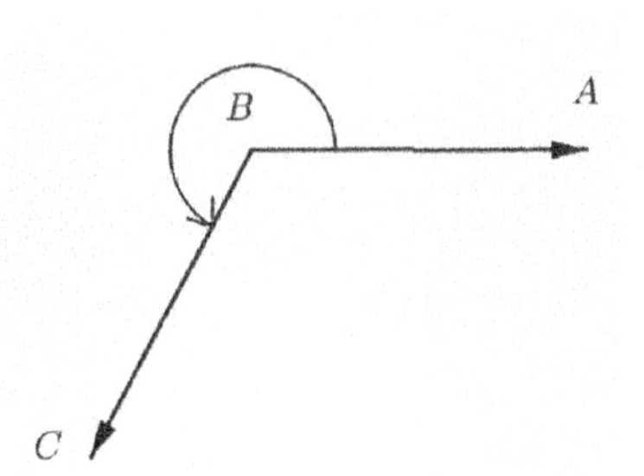

g.

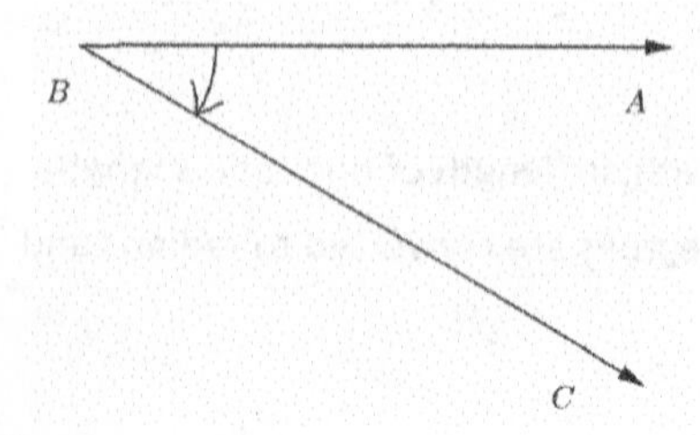

h.

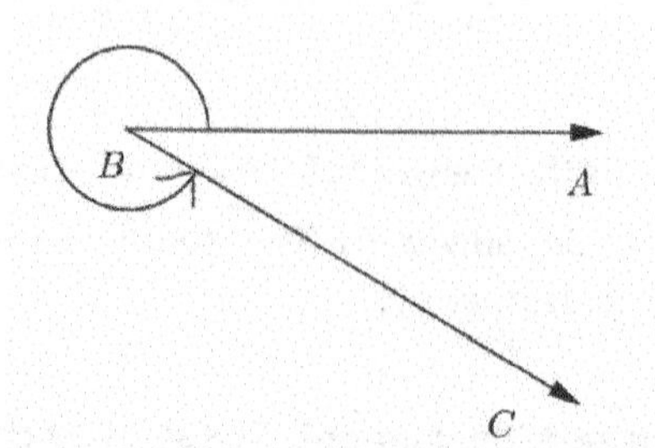

2. Complete the table below, converting from degrees to radians. Where appropriate, give your answers in the form of a fraction of π.

Degrees	Radians
$90°$	
$300°$	
$-45°$	
$-315°$	
$-690°$	
$3\frac{3}{4}°$	
$90\pi°$	
$-\frac{45°}{\pi}$	

3. Complete the table below, converting from radians to degrees.

Radians	Degrees
$\frac{\pi}{4}$	
$\frac{\pi}{6}$	
$\frac{5\pi}{12}$	
$\frac{11\pi}{36}$	
$-\frac{7\pi}{24}$	
$-\frac{11\pi}{12}$	
49π	
$\frac{49\pi}{3}$	

4. Use the unit circle diagram from the end of the lesson and your knowledge of the six trigonometric functions to complete the table below. Give your answers in exact form, as either rational numbers or radical expressions.

θ	$\cos(\theta)$	$\sin(\theta)$	$\tan(\theta)$	$\cot(\theta)$	$\sec(\theta)$	$\csc(\theta)$
$\frac{\pi}{3}$						
$\frac{3\pi}{4}$						
$\frac{5\pi}{6}$						
0						
$-\frac{3\pi}{4}$						
$-\frac{7\pi}{6}$						
$-\frac{11\pi}{3}$						

5. Use the unit circle diagram from the end of the lesson and your knowledge of the sine, cosine, and tangent functions to complete the table below. Select values of θ so that $0 \leq \theta < 2\pi$.

θ	$\cos(\theta)$	$\sin(\theta)$	$\tan(\theta)$
	$\frac{1}{2}$		$-\sqrt{3}$
		$-\frac{\sqrt{2}}{2}$	1
	$-\frac{\sqrt{2}}{2}$	$\frac{\sqrt{2}}{2}$	
	-1		0
	0	-1	
		$-\frac{1}{2}$	$\frac{\sqrt{3}}{3}$

6. How many radians does the minute hand of a clock rotate through over 10 minutes? How many degrees?

7. How many radians does the minute hand of a clock rotate through over half an hour? How many degrees?

8. What is the radian measure of an angle subtended by an arc of a circle with radius 4 cm if the intercepted arc has length 14 cm? How many degrees?

9. What is the radian measure of an angle formed by the minute and hour hands of a clock when the clock reads 1:30? How many degrees? (Hint: You must take into account that the hour hand is not directly on the 1.)

10. What is the radian measure of an angle formed by the minute and hour hands of a clock when the clock reads 5:45? How many degrees?

11. How many degrees does the earth revolve on its axis each hour? How many radians?

12. The distance from the equator to the North Pole is almost exactly $10{,}000$ km.
    a. Roughly how many kilometers is 1 degree of latitude?
    b. Roughly how many kilometers is 1 radian of latitude?

EUREKA MATH™

Lesson 10: Basic Trigonometric Identities from Graphs

Classwork

Exploratory Challenge 1

Consider the function $f(x) = \sin(x)$ where x is measured in radians.

Graph $f(x) = \sin(x)$ on the interval $[-\pi, 5\pi]$ by constructing a table of values. Include all intercepts, relative maximum points, and relative minimum points of the graph. Then, use the graph to answer the questions that follow.

x	
$f(x)$	

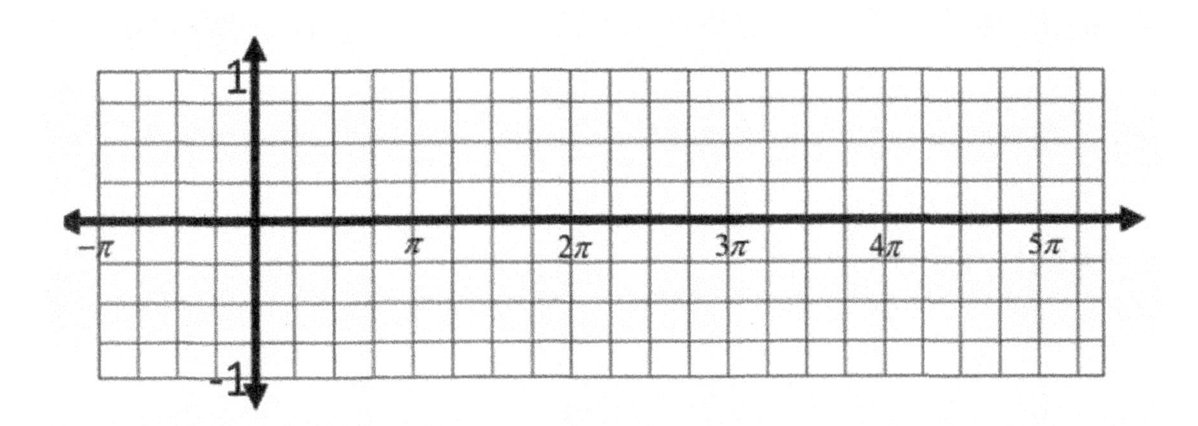

a. Using one of your colored pencils, mark the point on the graph at each of the following pairs of x-values.

$$x = -\frac{\pi}{2} \text{ and } x = -\frac{\pi}{2} + 2\pi$$

$$x = \pi \text{ and } x = \pi + 2\pi$$

$$x = \frac{7\pi}{4} \text{ and } x = \frac{7\pi}{4} + 2\pi$$

b. What can be said about the y-values for each pair of x-values marked on the graph?

c. Will this relationship hold for any two x-values that differ by 2π? Explain how you know.

d. Based on these results, make a conjecture by filling in the blank below.

For any real number x, $\sin(x + 2\pi) =$ ________________.

e. Test your conjecture by selecting another x-value from the graph and demonstrating that the equation from part (d) holds for that value of x.

f. How does the conjecture in part (d) support the claim that the sine function is a periodic function?

g. Use this identity to evaluate $\sin\left(\frac{13\pi}{6}\right)$.

h. Using one of your colored pencils, mark the point on the graph at each of the following pairs of x-values.

$$x = -\frac{\pi}{4} \text{ and } x = -\frac{\pi}{4} + \pi$$
$$x = 2\pi \text{ and } x = 2\pi + \pi$$
$$x = \frac{5\pi}{2} \text{ and } x = \frac{5\pi}{2} + \pi$$

i. What can be said about the y-values for each pair of x-values marked on the graph?

j. Will this relationship hold for any two x-values that differ by π? Explain how you know.

k. Based on these results, make a conjecture by filling in the blank below.

For any real number x, $\sin(x+\pi) =$ ________________.

l. Test your conjecture by selecting another x-value from the graph and demonstrating that the equation from part (k) holds for that value of x.

m. Is the following statement true or false? Use the conjecture from (k) to explain your answer.

$$\sin\left(\frac{4\pi}{3}\right) = -\sin\left(\frac{\pi}{3}\right)$$

n. Using one of your colored pencils, mark the point on the graph at each of the following pairs of x-values.

$$x = -\frac{3\pi}{4} \text{ and } x = \frac{3\pi}{4}$$
$$x = -\frac{\pi}{2} \text{ and } x = \frac{\pi}{2}$$

o. What can be said about the y-values for each pair of x-values marked on the graph?

p. Will this relationship hold for any two x-values with the same magnitude but opposite sign? Explain how you know.

q. Based on these results, make a conjecture by filling in the blank below.

For any real number x, $\sin(-x) =$ ________________.

r. Test your conjecture by selecting another x-value from the graph and demonstrating that the equation from part (q) holds for that value of x.

s. Is the sine function an odd function, even function, or neither? Use the identity from part (q) to explain.

t. Describe the x-intercepts of the graph of the sine function.

u. Describe the end behavior of the sine function.

Exploratory Challenge 2

Consider the function $g(x) = \cos(x)$ where x is measured in radians.

Graph $g(x) = \cos(x)$ on the interval $[-\pi, 5\pi]$ by constructing a table of values. Include all intercepts, relative maximum points, and relative minimum points. Then, use the graph to answer the questions that follow.

x	
$g(x)$	

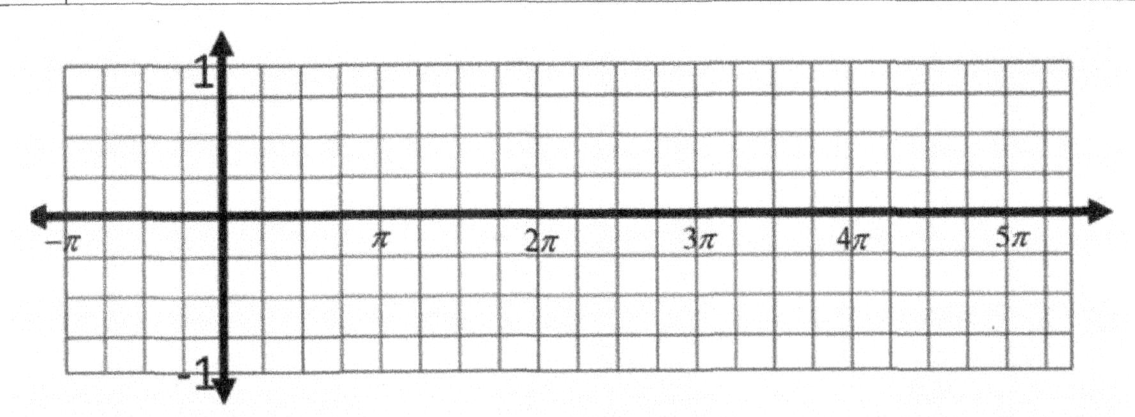

a. Using one of your colored pencils, mark the point on the graph at each of the following pairs of x-values.

$$x = -\frac{\pi}{2} \text{ and } x = -\frac{\pi}{2} + 2\pi$$

$$x = \pi \text{ and } x = \pi + 2\pi$$

$$x = \frac{7\pi}{4} \text{ and } x = \frac{7\pi}{4} + 2\pi$$

b. What can be said about the y-values for each pair of x-values marked on the graph?

c. Will this relationship hold for any two x-values that differ by 2π? Explain how you know.

d. Based on these results, make a conjecture by filling in the blank below.

For any real number x, $\cos(x + 2\pi) =$ ___________.

e. Test your conjecture by selecting another x-value from the graph and demonstrating that the equation from part (d) holds for that value of x.

f. How does the conjecture from part (d) support the claim that the cosine function is a periodic function?

g. Use this identity to evaluate $\cos\left(\frac{9\pi}{4}\right)$.

h. Using one of your colored pencils, mark the point on the graph at each of the following pairs of x-values.

$$x = -\frac{\pi}{4} \text{ and } x = -\frac{\pi}{4} + \pi$$

$$x = 2\pi \text{ and } x = 2\pi + \pi$$

$$x = \frac{5\pi}{2} \text{ and } x = \frac{5\pi}{2} + \pi$$

i. What can be said about the y-values for each pair of x-values marked on the graph?

j. Will this relationship hold for any two x-values that differ by π? Explain how you know.

k. Based on these results, make a conjecture by filling in the blank below.

For any real number x, $\cos(x + \pi) =$ __________.

l. Test your conjecture by selecting another x-value from the graph and demonstrating that the equation from part (k) holds for that value of x.

m. Is the following statement true or false? Use the identity from part (k) to explain your answer.

$$\cos\left(\frac{5\pi}{3}\right) = -\cos\left(\frac{2\pi}{3}\right)$$

n. Using one of your colored pencils, mark the point on the graph at each of the following pairs of x-values.

$$x = -\frac{3\pi}{4} \text{ and } x = \frac{3\pi}{4}$$

$$x = -\pi \text{ and } x = \pi$$

o. What can be said about the y-values for each pair of x-values marked on the graph?

p. Will this relationship hold for any two x-values with the same magnitude and same sign? Explain how you know.

q. Based on these results, make a conjecture by filling in the blank below.

For any real number, $\cos(-x) =$ ____________.

r. Test your conjecture by selecting another x-value from the graph and demonstrating that the identity is true for that value of x.

s. Is the cosine function an odd function, even function, or neither? Use the identity from part (n) to explain.

t. Describe the x-intercepts of the graph of the cosine function.

u. Describe the end behavior of $g(x) = \cos(x)$.

Exploratory Challenge 3

Graph both $f(x) = \sin(x)$ and $g(x) = \cos(x)$ on the graph below. Then, use the graphs to answer the questions that follow.

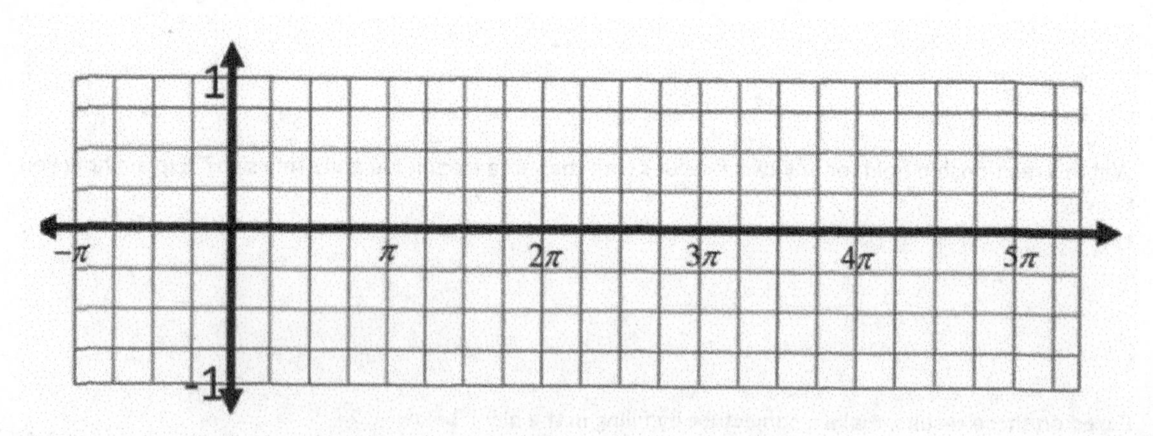

a. List ways in which the graphs of the sine and cosine functions are alike.

b. List ways in which the graphs of the sine and cosine functions are different.

c. What type of transformation would be required to make the graph of the sine function coincide with the graph of the cosine function?

d. What is the smallest possible horizontal translation required to make the graph of $f(x) = \sin(x)$ coincide with the graph of $g(x) = \cos(x)$?

e. What is the smallest possible horizontal translation required to make the graph of $g(x) = \cos(x)$ coincide with the graph of $f(x) = \sin(x)$?

f. Use your answers from parts (d) and (e) to fill in the blank below.

For any real number x, ____________ $= \cos\left(x - \frac{\pi}{2}\right)$.

For any real number x, ____________ $= \sin\left(x + \frac{\pi}{2}\right)$.

Lesson Summary

For all real numbers x:

$$\sin(x + 2\pi) = \sin(x) \qquad \cos(x + 2\pi) = \cos(x)$$
$$\sin(x + \pi) = -\sin(x) \qquad \cos(x + \pi) = -\cos(x)$$
$$\sin(-x) = -\sin(x) \qquad \cos(-x) = -\cos(x)$$
$$\sin\left(x + \frac{\pi}{2}\right) = \cos(x) \qquad \cos\left(x - \frac{\pi}{2}\right) = \sin(x)$$

Problem Set

1. Describe the values of x for which each of the following is true.
   a. The cosine function has a relative maximum.
   b. The sine function has a relative maximum.

2. Without using a calculator, rewrite each of the following in order from least to greatest. Use the graph to explain your reasoning.

   $\sin\left(\frac{\pi}{4}\right)$ $\quad$ $\sin\left(-\frac{2\pi}{3}\right)$ $\quad$ $\sin\left(\frac{\pi}{4}\right)$ $\quad$ $\sin\left(-\frac{\pi}{2}\right)$

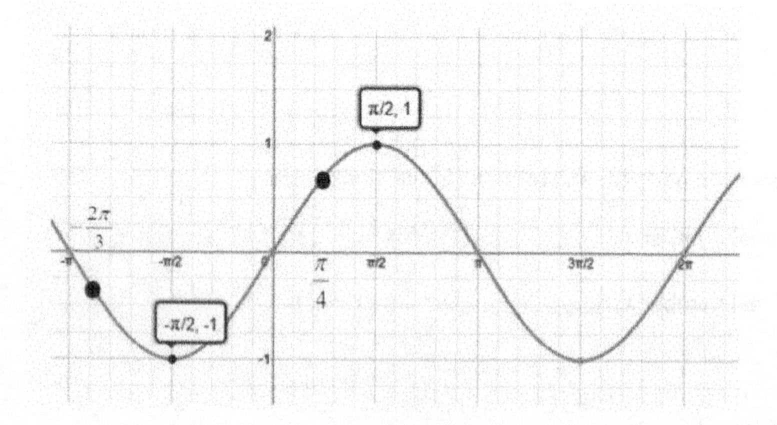

EUREKA MATH

3. Without using a calculator, rewrite each of the following in order from least to greatest. Use the graph to explain your reasoning.

$\cos\left(\frac{\pi}{2}\right)$ $\cos\left(\frac{5\pi}{4}\right)$ $\cos\left(\frac{\pi}{4}\right)$ $\cos(5\pi)$

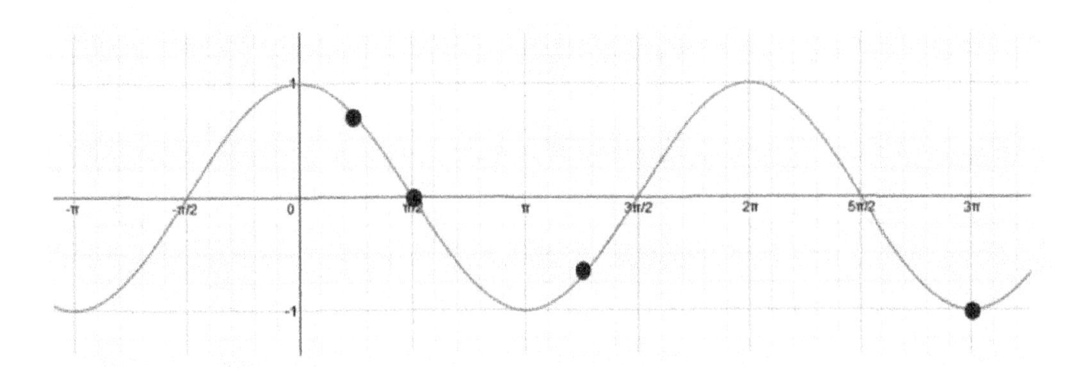

4. Evaluate each of the following without a calculator using a trigonometric identity when needed.

$\cos\left(\frac{\pi}{6}\right)$ $\cos\left(-\frac{\pi}{6}\right)$ $\cos\left(\frac{7\pi}{6}\right)$ $\cos\left(\frac{13\pi}{6}\right)$

5. Evaluate each of the following without a calculator using a trigonometric identity when needed.

$\sin\left(\frac{3\pi}{4}\right)$ $\sin\left(\frac{11\pi}{4}\right)$ $\sin\left(\frac{7\pi}{4}\right)$ $\sin\left(\frac{-5\pi}{4}\right)$

6. Use the rotation through θ radians shown to answer each of the following questions.
   a. Explain why $\sin(-\theta) = -\sin(\theta)$ for all real numbers θ.
   b. What symmetry does this identity demonstrate about the graph of $y = \sin(x)$?

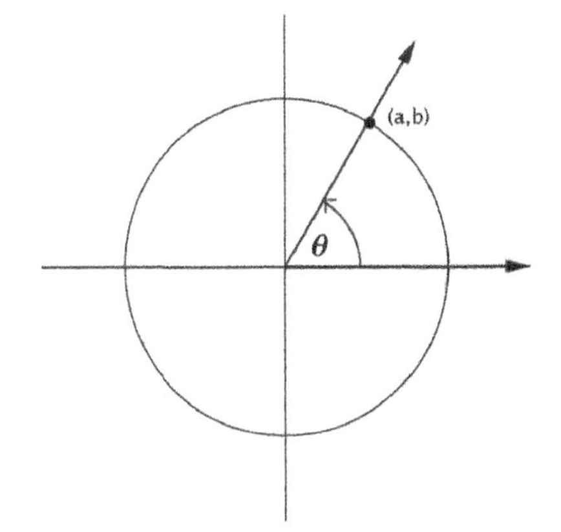

7. Use the same rotation shown in Problem 6 to answer each of the following questions.
   a. Explain why $\cos(-\theta) = \cos(\theta)$.
   b. What symmetry does this identity demonstrate about the graph of $y = \cos(x)$?

8. Find equations of two different functions that can be represented by the graph shown below—one sine and one cosine—using different horizontal transformations.

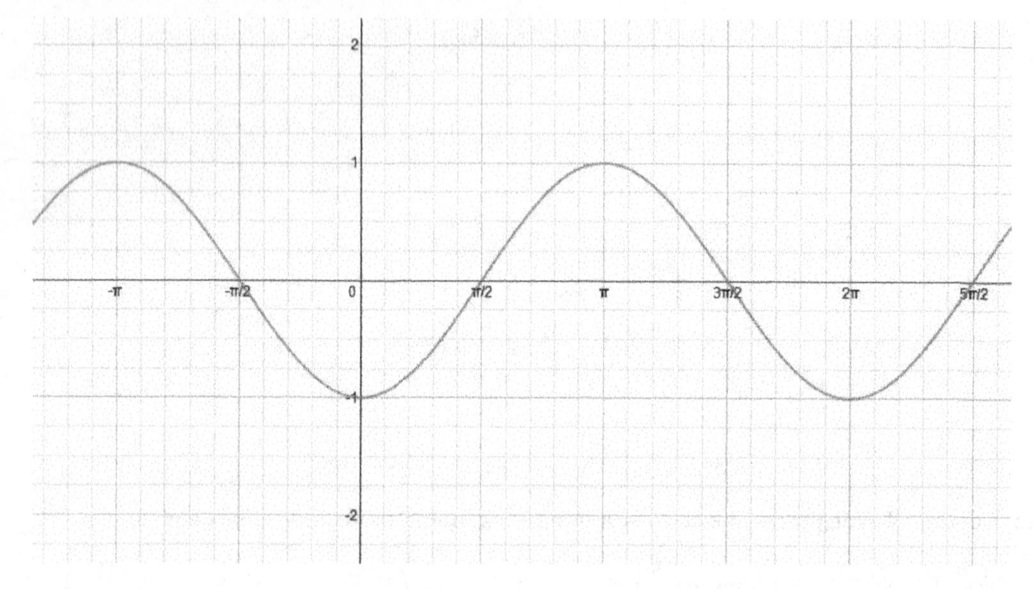

9. Find equations of two different functions that can be represented by the graph shown below—one sine and one cosine—using different horizontal translations.

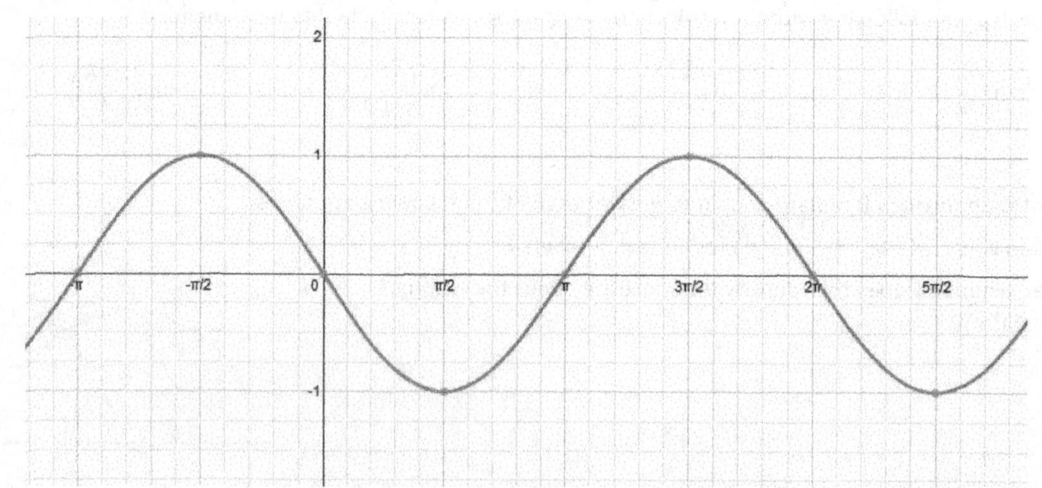

Lesson 11: Transforming the Graph of the Sine Function

Classwork

Opening Exercise

Explore your assigned parameter in the sinusoidal function $f(x) = A\sin(\omega(x-h)) + k$. Select several different values for your assigned parameter, and explore the effects of changing the parameter's value on the graph of the function compared to the graph of $f(x) = \sin(x)$. Record your observations in the table below. Include written descriptions and sketches of graphs.

A-Team	ω-Team
$f(x) = A\sin(x)$ Suggested A values: $2, 3, 10, 0, -1, -2, \frac{1}{2}, \frac{1}{5}, -\frac{1}{3}$	$f(x) = \sin(\omega x)$ Suggested ω values: $2, 3, 5, \frac{1}{2}, \frac{1}{4}, 0, -1, -2, \pi, 2\pi, 3\pi, \frac{\pi}{2}, \frac{\pi}{4}$

k-Team	**h-Team**
$f(x) = \sin(x) + k$	$f(x) = \sin(x - h)$
Suggested k values: $2, 3, 10, 0, -1, -2, \frac{1}{2}, \frac{1}{5}, -\frac{1}{3}$	Suggested h values: $\pi, -\pi, \frac{\pi}{2}, -\frac{\pi}{4}, 2\pi, 2, 0, -1, -2, 5, -5$

Example

Graph the following function:

$$f(x) = 3\sin\left(4\left(x - \frac{\pi}{6}\right)\right) + 2.$$

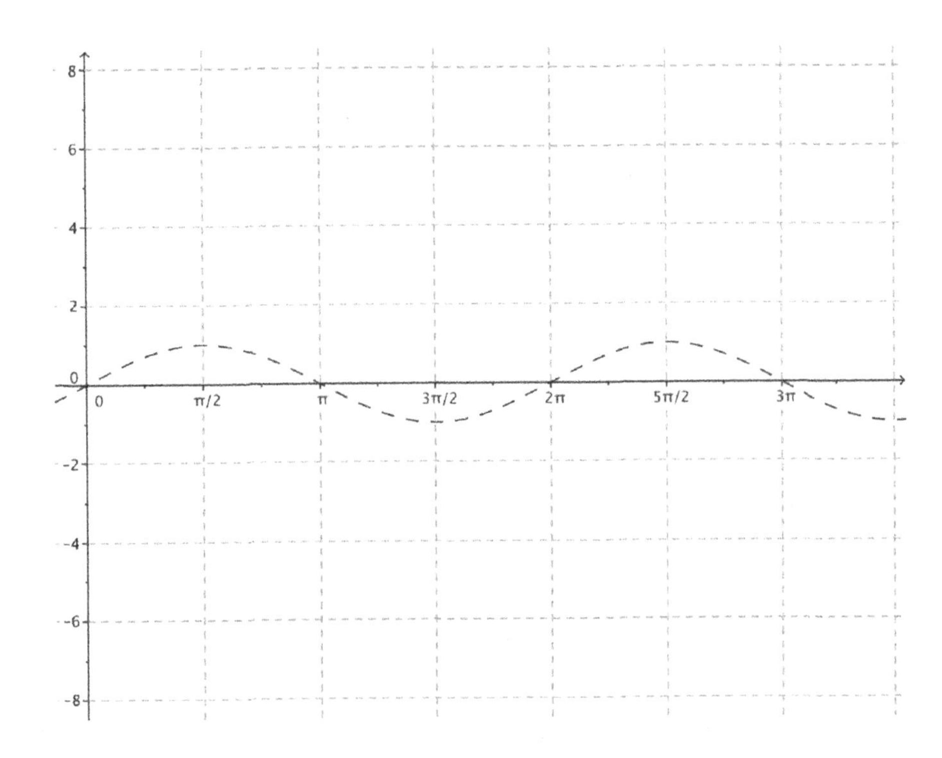

Exercise

For each function, indicate the amplitude, frequency, period, phase shift, vertical translation, and equation of the midline. Graph the function together with a graph of the sine function $f(x) = \sin(x)$ on the same axes. Graph at least one full period of each function.

a. $g(x) = 3\sin(2x) - 1$

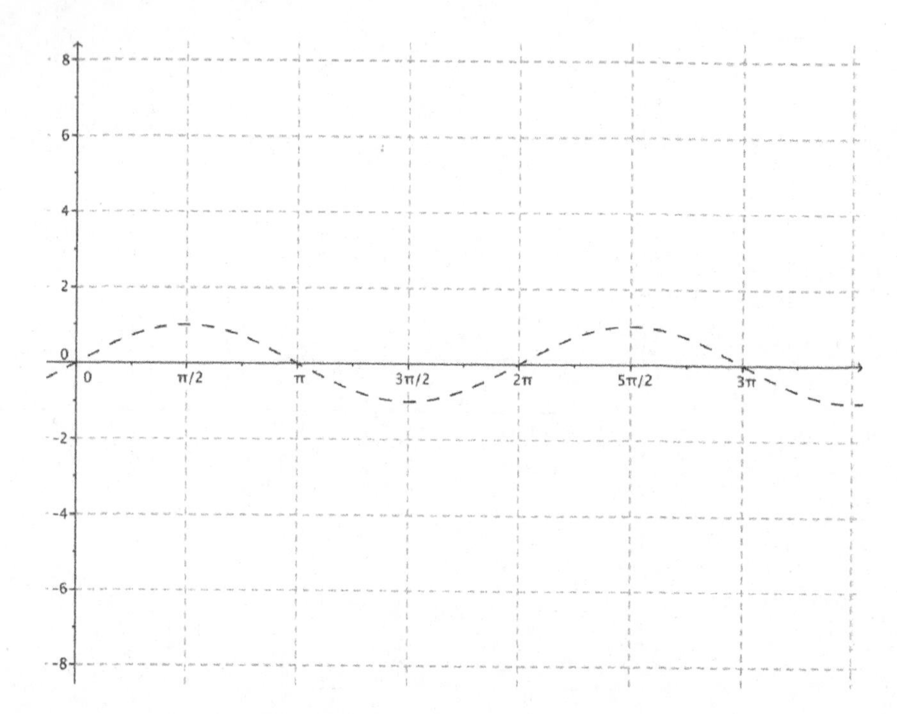

b. $g(x) = \frac{1}{2}\sin\left(\frac{1}{4}(x + \pi)\right)$

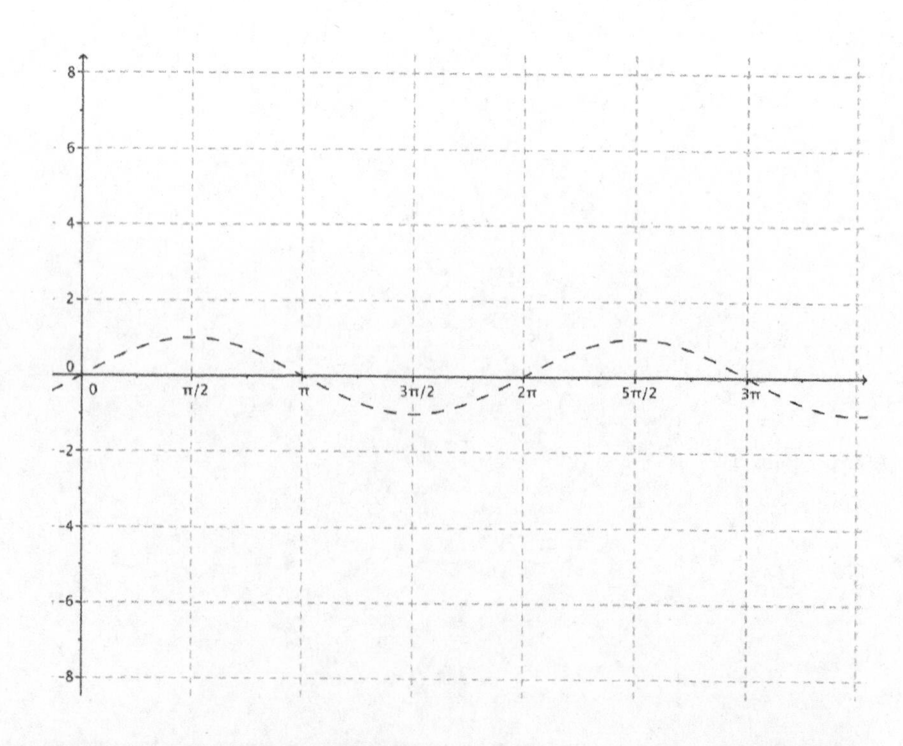

c. $g(x) = 5\sin(-2x) + 2$

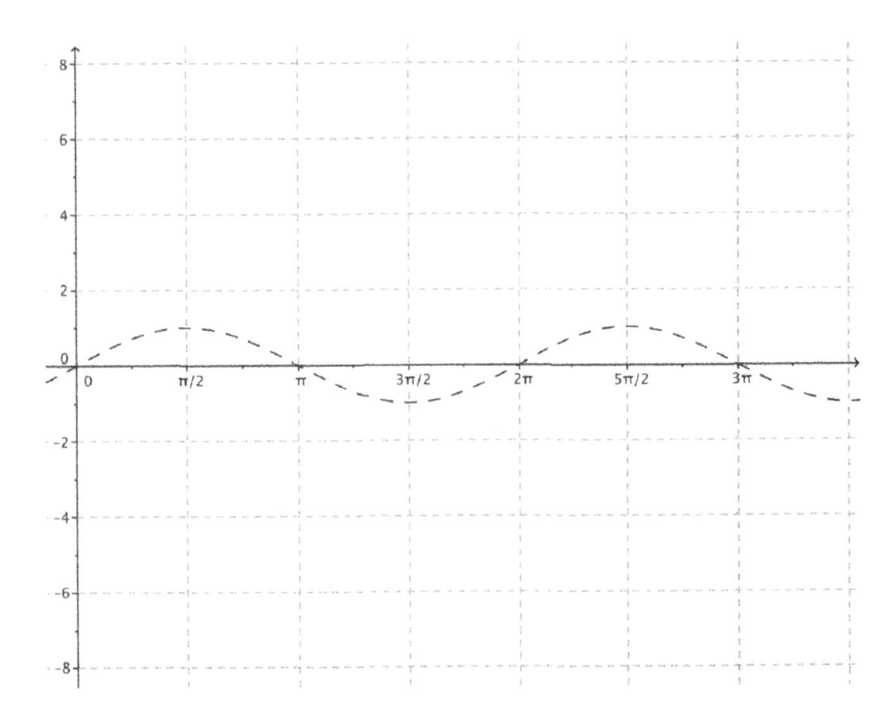

d. $g(x) = -2\sin\left(3\left(x + \frac{\pi}{6}\right)\right)$

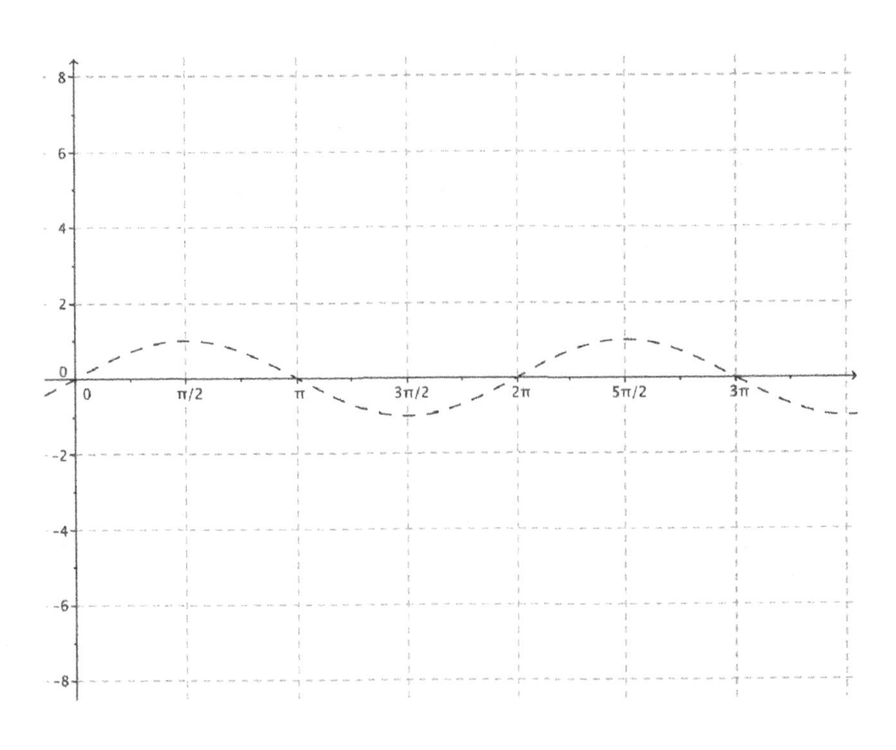

e. $g(x) = 3\sin(x + \pi) + 3$

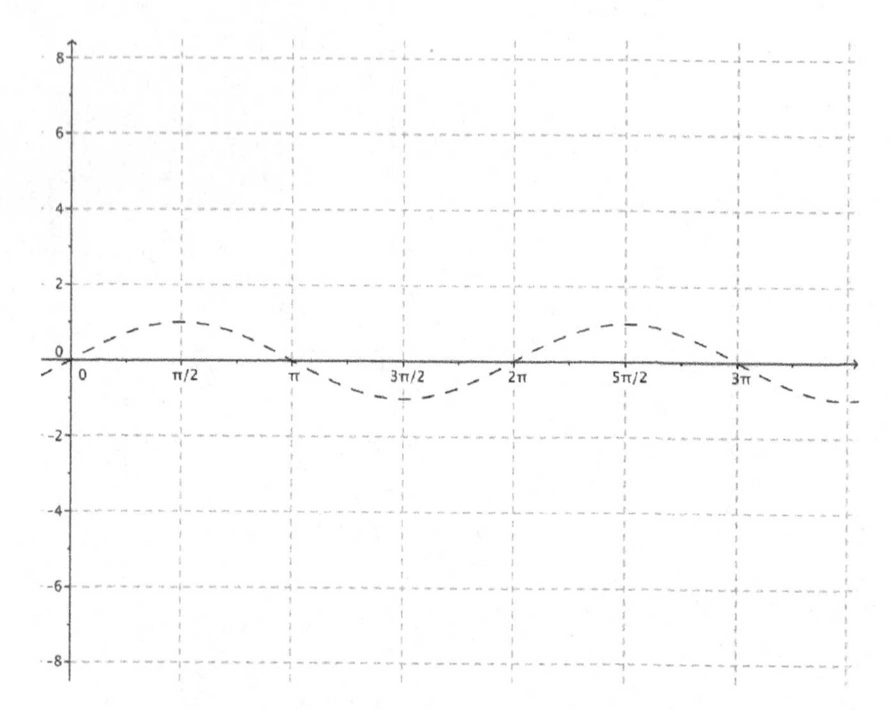

f. $g(x) = -\frac{2}{3}\sin(4x) - 3$

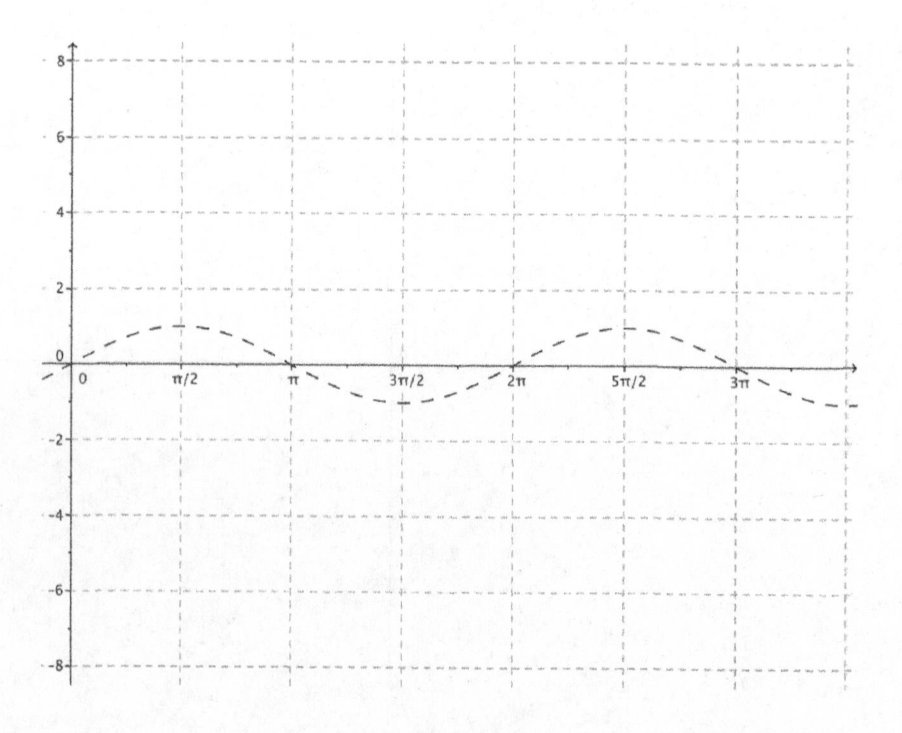

g. $g(x) = \pi \sin\left(\frac{x}{2}\right) + \pi$

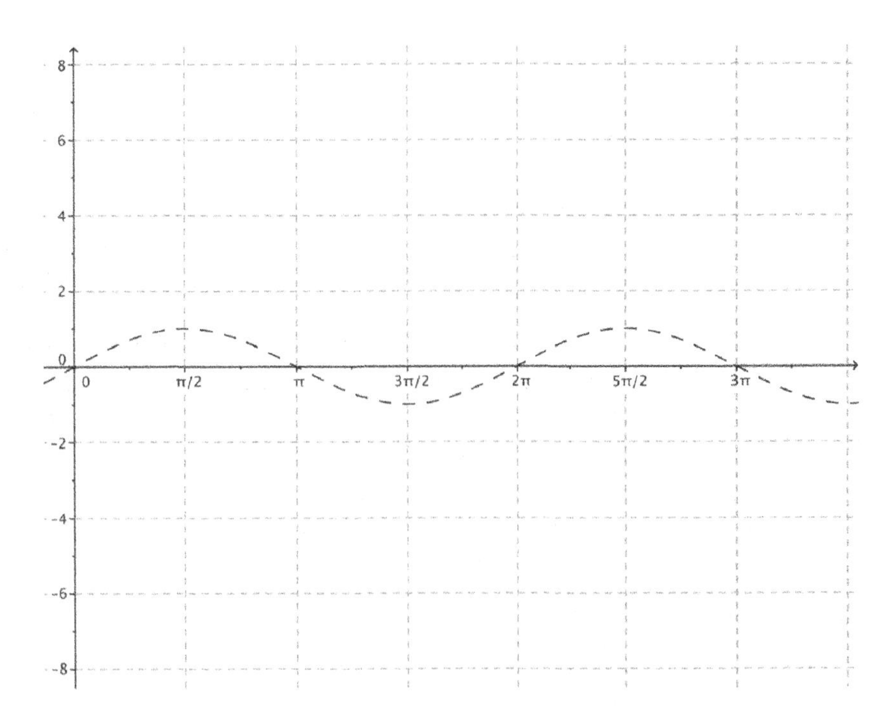

h. $g(x) = 4\sin\left(\frac{1}{2}(x - 5\pi)\right)$

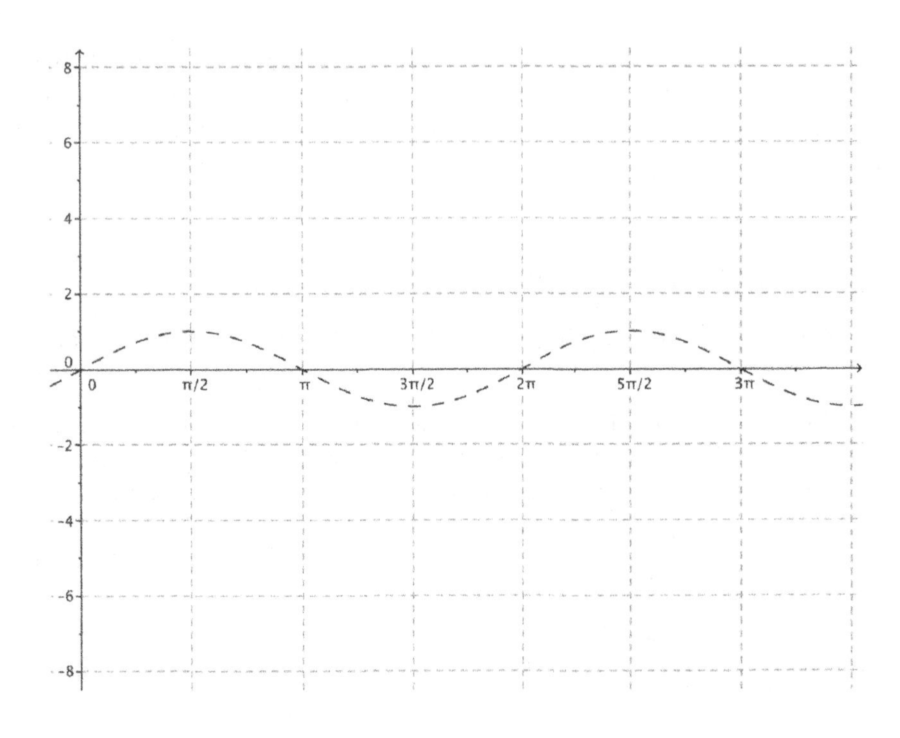

Lesson Summary

This lesson investigated the effects of the parameters A, ω, h, and k on the graph of the function $f(x) = A\sin(\omega(x-h)) + k$.

- The amplitude of the function is $|A|$; the vertical distance from a maximum point to the midline of the graph is $|A|$.
- The frequency of the function is $f = \frac{|\omega|}{2\pi}$, and the period is $P = \frac{2\pi}{|\omega|}$. The period is the vertical distance between two consecutive maximal points on the graph of the function.
- The phase shift is h. The value of h determines the horizontal translation of the graph from the graph of the sine function. If $h > 0$, the graph is translated h units to the right, and if $h < 0$, the graph is translated h units to the left.
- The graph of $y = k$ is the midline. The value of k determines the vertical translation of the graph compared to the graph of the sine function. If $k > 0$, then the graph shifts k units upward. If $k < 0$, then the graph shifts k units downward.

These parameters affect the graph of $f(x) = A\cos(\omega(x-h)) + k$ similarly.

Problem Set

1. For each function, indicate the amplitude, frequency, period, phase shift, horizontal and vertical translations, and equation of the midline. Graph the function together with a graph of the sine function $f(x) = \sin(x)$ on the same axes. Graph at least one full period of each function. No calculators are allowed.
   a. $g(x) = 3\sin\left(x - \frac{\pi}{4}\right)$
   b. $g(x) = 5\sin(4x)$
   c. $g(x) = 4\sin\left(3\left(x + \frac{\pi}{2}\right)\right)$
   d. $g(x) = 6\sin(2x + 3\pi)$ (Hint: First, rewrite the function in the form $g(x) = A\sin(\omega(x-h))$.)

2. For each function, indicate the amplitude, frequency, period, phase shift, horizontal and vertical translations, and equation of the midline. Graph the function together with a graph of the sine function $f(x) = \cos(x)$ on the same axes. Graph at least one full period of each function. No calculators are allowed.
   a. $g(x) = \cos(3x)$
   b. $g(x) = \cos\left(x - \frac{3\pi}{4}\right)$
   c. $g(x) = 3\cos\left(\frac{x}{4}\right)$
   d. $g(x) = 3\cos(2x) - 4$
   e. $g(x) = 4\cos\left(\frac{\pi}{4} - 2x\right)$ (Hint: First, rewrite the function in the form $g(x) = A\cos(\omega(x-h))$.)

3. For each problem, sketch the graph of the pairs of indicated functions on the same set of axes without using a calculator or other graphing technology.
   a. $f(x) = \sin(4x)$, $g(x) = \sin(4x) + 2$
   b. $f(x) = \sin\left(\frac{1}{2}x\right)$, $g(x) = 3\sin\left(\frac{1}{2}x\right)$
   c. $f(x) = \sin(-2x)$, $g(x) = \sin(-2x) - 3$
   d. $f(x) = 3\sin(x)$, $g(x) = 3\sin\left(x - \frac{\pi}{2}\right)$
   e. $f(x) = -4\sin(x)$, $g(x) = -4\sin\left(\frac{1}{3}x\right)$
   f. $f(x) = \frac{3}{4}\sin(x)$, $g(x) = \frac{3}{4}\sin(x - 1)$
   g. $f(x) = \sin(2x)$, $g(x) = \sin\left(2\left(x - \frac{\pi}{6}\right)\right)$
   h. $f(x) = 4\sin(x) - 3$, $g(x) = 4\sin\left(x - \frac{\pi}{4}\right) - 3$

Extension:

4. Show that if the graphs of the functions $f(x) = A\sin\big(\omega(x - h_1)\big) + k$ and $g(x) = A\sin\big(\omega(x - h_2)\big) + k$ are the same, then h_1 and h_2 differ by an integer multiple of the period.

5. Show that if h_1 and h_2 differ by an integer multiple of the period, then the graphs of $f(x) = A\sin\big(\omega(x - h_1)\big) + k$ and $g(x) = A\sin\big(\omega(x - h_2)\big) + k$ are the same graph.

6. Find the x-intercepts of the graph of the function $f(x) = A\sin\big(\omega(x - h)\big)$ in terms of the period P, where $\omega > 0$.

This page intentionally left blank

Lesson 12: Ferris Wheels—Using Trigonometric Functions to Model Cyclical Behavior

Classwork

Opening Exercise

Ernesto thinks that this is the graph of $f(x) = 4\sin\left(x - \frac{\pi}{2}\right)$. Danielle thinks it is the graph of $f(x) = 4\cos(x)$.

Who is correct and why?

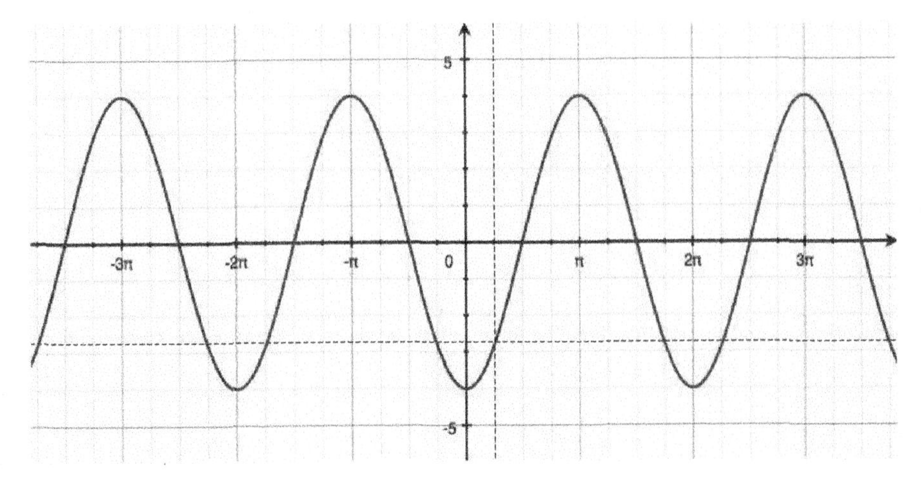

Exploratory Challenge/Exercises 1–5

A carnival has a Ferris wheel that is 50 feet in diameter with 12 passenger cars. When viewed from the side where passengers board, the Ferris wheel rotates counterclockwise and makes two full turns each minute. Riders board the Ferris wheel from a platform that is 15 feet above the ground. We will use what we have learned about periodic functions to model the position of the passenger cars from different mathematical perspectives. We will use the points on the circle in the diagram on the right to represent the position of the cars on the wheel.

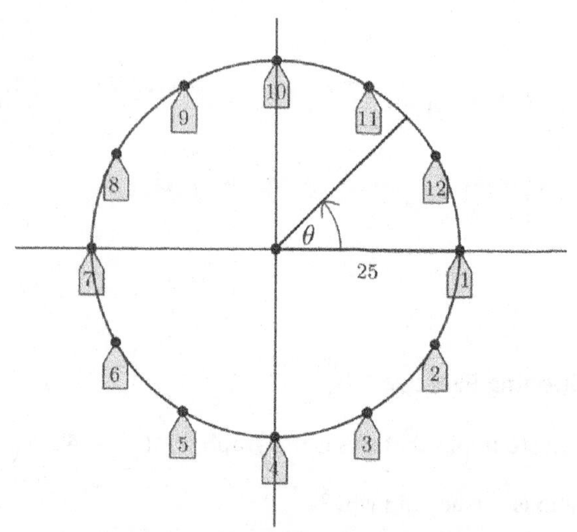

1. For this exercise, we will consider the height of a passenger car to be the vertical displacement from the horizontal line through the center of the wheel and the co-height of a passenger car to be the horizontal displacement from the vertical line through the center of the wheel.

   a. Let $\theta = 0$ represent the position of car 1 in the diagram above. Sketch the graphs of the co-height and the height of car 1 as functions of θ, the number of radians through which the car has rotated.

   b. What is the amplitude, $|A|$, of the height and co-height functions for this Ferris wheel?

c. Let $X(\theta)$ represent the co-height function after rotation by θ radians, and let $Y(\theta)$ represent the height function after rotation by θ radians. Write functions X for the co-height and Y for the height in terms of θ.

d. Graph the functions X and Y from part (c) on a graphing calculator set to parametric mode. Use a viewing window $[-48, 48] \times [-30, 30]$. Sketch the graph below.

e. Why did we choose the symbols X and Y to represent the co-height and height functions?

f. Evaluate $X(0)$ and $Y(0)$, and explain their meaning in the context of the Ferris wheel.

g. Evaluate $X\left(\frac{\pi}{2}\right)$ and $Y\left(\frac{\pi}{2}\right)$, and explain their meaning in the context of the Ferris wheel.

2. The model we created in Exercise 1 measures the height of car 1 above the horizontal line through the center of the wheel. We now want to alter this model so that it measures the height of car 1 above the ground.

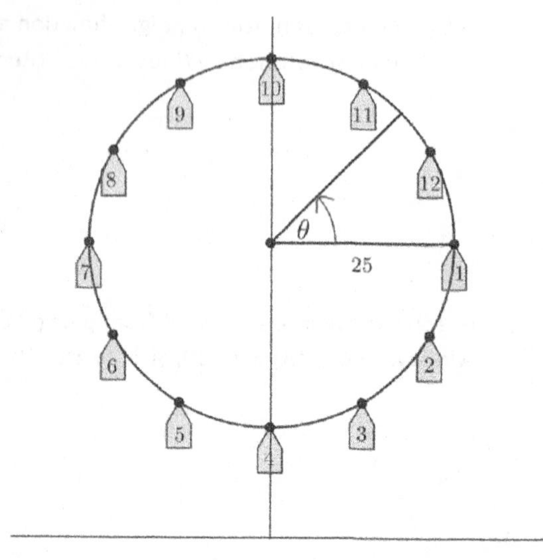

   a. If we measure the height of car 1 above the ground instead of above the horizontal line through the center of the wheel, how will the functions X and Y need to change?

   b. Let $\theta = 0$ represent the position of car 1 in the diagram to the right. Sketch the graphs of the co-height and the height of car 1 as functions of the number of radians through which the car has rotated, θ.

   c. How are the graphs from part (b) related to the graphs from Exercise 1(a)?

d. From this perspective, find the equations for the functions X and Y that model the position of car 1 with respect to the number of radians the car has rotated.

e. Change the viewing window on your calculator to $[-60, 60] \times [-5, 70]$, and graph the functions X and Y together. Sketch the graph.

f. Evaluate $X(0)$ and $Y(0)$, and explain their meaning in the context of the Ferris wheel.

g. Evaluate $X\left(\frac{\pi}{2}\right)$ and $Y\left(\frac{\pi}{2}\right)$, and explain their meaning in the context of the Ferris wheel.

3. In reality, no one boards a Ferris wheel halfway up; passengers board at the bottom of the wheel. To truly model the motion of a Ferris wheel, we need to start with passengers on the bottom of the wheel. Refer to the diagram below.

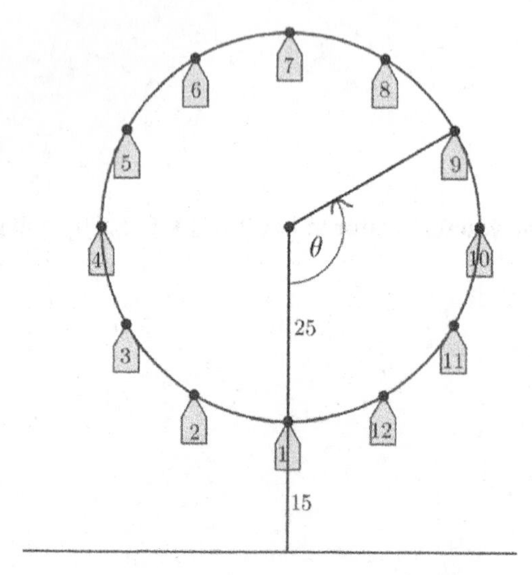

a. Let $\theta = 0$ represent the position of car 1 at the bottom of the wheel in the diagram above. Sketch the graphs of the height and the co-height of car 1 as functions of θ, the number of radians through which the car has rotated from the position at the bottom of the wheel.

b. How are the graphs from part (a) related to the graphs from Exercise 2(b)?

c. From this perspective, find the equations for the functions X and Y that model the position of car 1 with respect to the number of radians the car has rotated.

d. Graph the functions X and Y from part (c) together on the graphing calculator. Sketch the graph. How is this graph different from the one from Exercise 2(e)?

e. Evaluate $X(0)$ and $Y(0)$, and explain their meaning in the context of the Ferris wheel.

f. Evaluate $X\left(\frac{\pi}{2}\right)$ and $Y\left(\frac{\pi}{2}\right)$, and explain their meaning in the context of the Ferris wheel.

4. Finally, it is not very useful to track the position of a Ferris wheel as a function of how much it has rotated. It would make more sense to keep track of the Ferris wheel as a function of time. Recall that the Ferris wheel completes two full turns per minute.

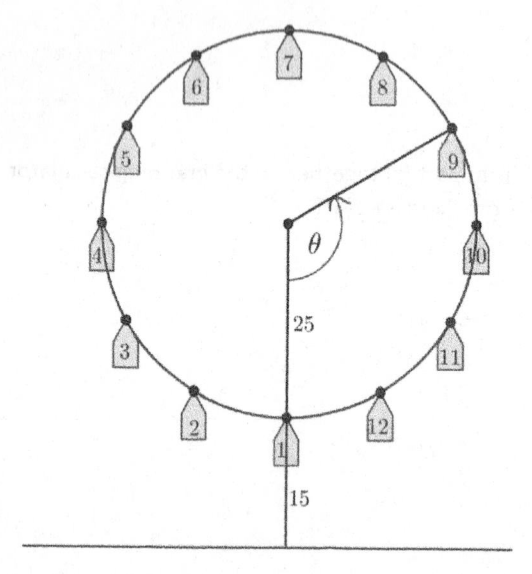

a. Let $\theta = 0$ represent the position of car 1 at the bottom of the wheel. Sketch the graphs of the co-height and the height of car 1 as functions of time.

b. The co-height and height functions from part (a) can be written in the form $X(t) = A\cos(\omega(t-h)) + k$ and $Y(t) = A\sin(\omega(t-h)) + k$. From the graphs in part (a), identify the values of A, ω, h, and k for each function X and Y.

c. Write the equations $X(t)$ and $Y(t)$ using the values you identified in part (b).

d. In function mode, graph your functions from part (c) on a graphing calculator for $0 < t < 2$, and compare against your sketches in part (a) to confirm your equations.

e. Explain the meaning of the parameters in your equation for X in terms of the Ferris wheel scenario.

f. Explain the meaning of the parameters in your equation for Y in terms of the Ferris wheel scenario.

5. In parametric mode, graph the functions X and Y from Exercise 3(c) on a graphing calculator for $0 \le t \le \frac{1}{2}$.

   a. Sketch the graph. How is this graph different from the graph in Exercise 3(d)?

   b. What would the graph look like if you graphed the functions X and Y from Exercise 3(c) for $0 \le t \le \frac{1}{4}$? Why?

c. Evaluate $X(0)$ and $Y(0)$, and explain their meaning in the context of the Ferris wheel.

d. Evaluate $X\left(\frac{1}{8}\right)$ and $Y\left(\frac{1}{8}\right)$, and explain their meaning in the context of the Ferris wheel.

Exercise 6–9

6. You are in car 1, and you see your best friend in car 3. How much higher than your friend are you when you reach the top?

7. Find an equation of the function H that represents the difference in height between you in car 1 and your friend in car 3 as the wheel rotates through θ radians, beginning with $\theta = 0$ at the bottom of the wheel.

8. Find an equation of the function that represents the difference in height between car 1 and car 3 with respect to time, t, in minutes. Let $t = 0$ minutes correspond to a time when car 1 is located at the bottom of the Ferris wheel. Assume the wheel is moving at a constant speed starting at $t = 0$.

9. Use a calculator to graph the function H in Exercise 8 for $0 \leq t \leq 2$. What type of function does this appear to be? Does that make sense?

Problem Set

1. In the classic novel *Don Quixote*, the title character famously battles a windmill. In this problem, you will model what happens when Don Quixote battles a windmill, and the windmill wins. Suppose the center of the windmill is 20 feet off the ground, and the sails are 15 feet long. Don Quixote is caught on a tip of one of the sails. The sails are turning at a rate of one counterclockwise rotation every 60 seconds.

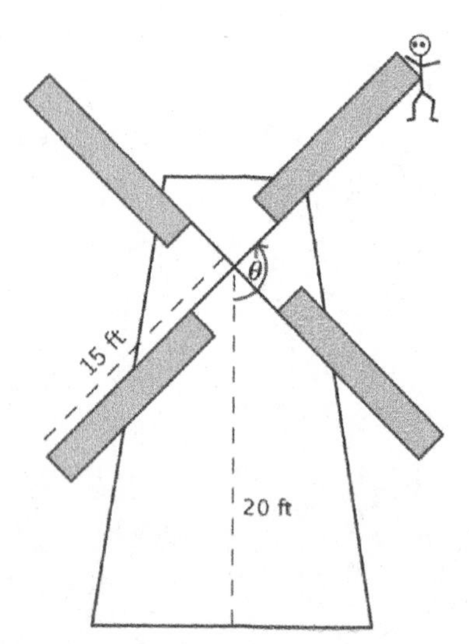

   a. Explain why a sinusoidal function could be used to model Don Quixote's height above the ground as a function of time.
   b. Sketch a graph of Don Quixote's height above the ground as a function of time. Assume $t = 0$ corresponds to a time when he was closest to the ground. What are the amplitude, period, and midline of the graph?
   c. Model Don Quixote's height H above the ground as a function of time t since he was closest to the ground.
   d. After 1 minute and 40 seconds, Don Quixote fell off the sail and straight down to the ground. How far did he fall?

2. The High Roller, a Ferris wheel in Las Vegas, Nevada, opened in March 2014. The 550 ft. tall wheel has a diameter of 520 feet. A ride on one of its 28 passenger cars lasts 30 minutes, the time it takes the wheel to complete one full rotation. Riders board the passenger cars at the bottom of the wheel. Assume that once the wheel is in motion, it maintains a constant speed for the 30-minute ride and is rotating in a counterclockwise direction.
   a. Sketch a graph of the height of a passenger car on the High Roller as a function of the time the ride began.
   b. Write a sinusoidal function H that represents the height of a passenger car t minutes after the ride begins.
   c. Explain how the parameters of your sinusoidal function relate to the situation.
   d. If you were on this ride, how high would you be above the ground after 20 minutes?
   e. Suppose the ride costs $25. How much does 1 minute of riding time cost? How much does 1 foot of riding distance cost? How much does 1 foot of height above the ground cost?
   f. What are some of the limitations of this model based on the assumptions that we made?

3. Once in motion, a pendulum's distance varies sinusoidally from 12 feet to 2 feet away from a wall every 12 seconds.
   a. Sketch the pendulum's distance D from the wall over a 1-minute interval as a function of time t. Assume $t = 0$ corresponds to a time when the pendulum was furthest from the wall.
   b. Write a sinusoidal function for D, the pendulum's distance from the wall, as a function of the time since it was furthest from the wall.
   c. Identify two different times when the pendulum was 10 feet away from the wall. (Hint: Write an equation, and solve it graphically.)

4. The height in meters relative to the starting platform height of a car on a portion of a roller coaster track is modeled by the function $H(t) = 3\sin\left(\frac{\pi}{4}(t - 10)\right) - 7$. It takes a car 24 seconds to travel on this portion of the track, which starts 10 seconds into the ride.
   a. Graph the height relative to the starting platform as a function of time over this time interval.
   b. Explain the meaning of each parameter in the height function in terms of the situation.

5. Given the following function, use the parameters to formulate a real-world situation involving one dimension of circular motion that could be modeled using this function. Explain how each parameter of the function relates to your situation.

$$f(x) = 10\sin\left(\frac{\pi}{8}(x - 3)\right) + 15$$

This page intentionally left blank

Lesson 13: Tides, Sound Waves, and Stock Markets

Classwork

Opening Exercise

Anyone who works on or around the ocean needs to have information about the changing tides to conduct their business safely and effectively. People who go to the beach or out onto the ocean for recreational purposes also need information about tides when planning their trip. The table below shows tide data for Montauk, NY, for May 21–22, 2014. The heights reported in the table are relative to the Mean Lower Low Water (MLLW). The MLLW is the average height of the lowest tide recorded at a tide station each day during the recording period. This reference point is used by the National Oceanic and Atmospheric Administration (NOAA) for the purposes of reporting tidal data throughout the United States. Each different tide station throughout the United States has its own MLLW. High and low tide levels are reported relative to this number. Since it is an average, some low tide values can be negative. NOAA resets the MLLW values approximately every 20 years.

MONTAUK, NY, TIDE CHART

Date	Day	Time	Height in Feet	High/Low
2014/05/21	Wed.	02:47 a.m.	2.48	H
2014/05/21	Wed.	09:46 a.m.	-0.02	L
2014/05/21	Wed.	03:31 p.m.	2.39	H
2014/05/21	Wed.	10:20 p.m.	0.27	L
2014/05/22	Thurs.	03:50 a.m.	2.30	H
2014/05/22	Thurs.	10:41 a.m.	0.02	L
2014/05/22	Thurs.	04:35 p.m.	2.51	H
2014/05/22	Thurs.	11:23 p.m.	0.21	L

a. Create a scatter plot of the data with the horizontal axis representing time since midnight on May 21 and the vertical axis representing the height in feet relative to the MLLW.

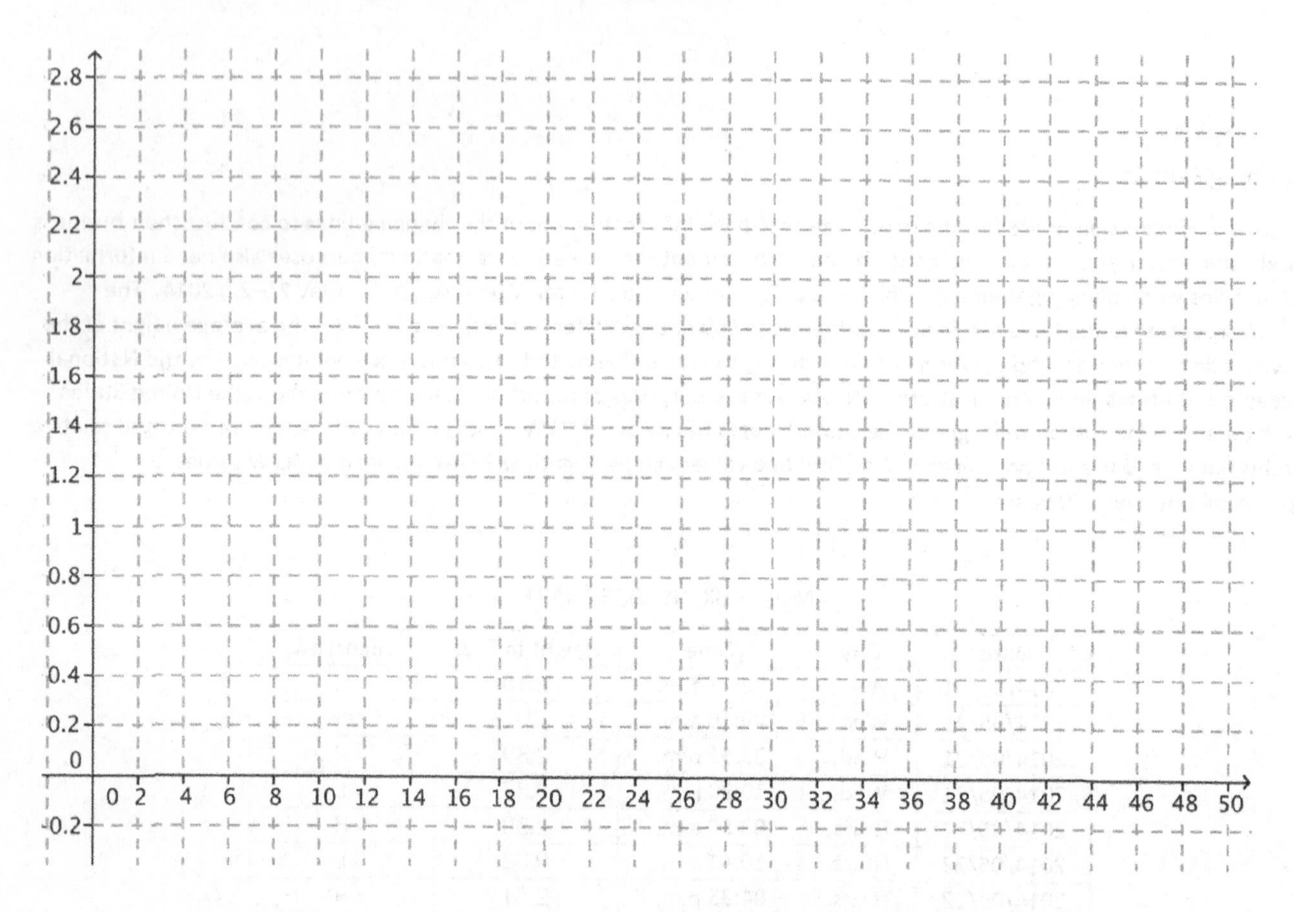

b. What type of function would best model this set of data? Explain your choice.

Example 1: Write a Sinusoidal Function to Model a Set of Data

a. On the scatter plot you created in the Opening Exercise, sketch the graph of a sinusoidal function that could be used to model the data.

b. What are the midline, amplitude, and period of the graph?

c. Estimate the horizontal distance between a point on the graph that crosses the midline and the vertical axis.

d. Write a function of the form $f(x) = A\sin(\omega(x-h)) + k$ to model these data, where x is the hours since midnight on May 21, and $f(x)$ is the height in feet relative to the MLLW.

Exercise 1

1. The graph of the tides at Montauk for the week of May 21–28 is shown below. How accurately does your model predict the time and height of the high tide on May 28?

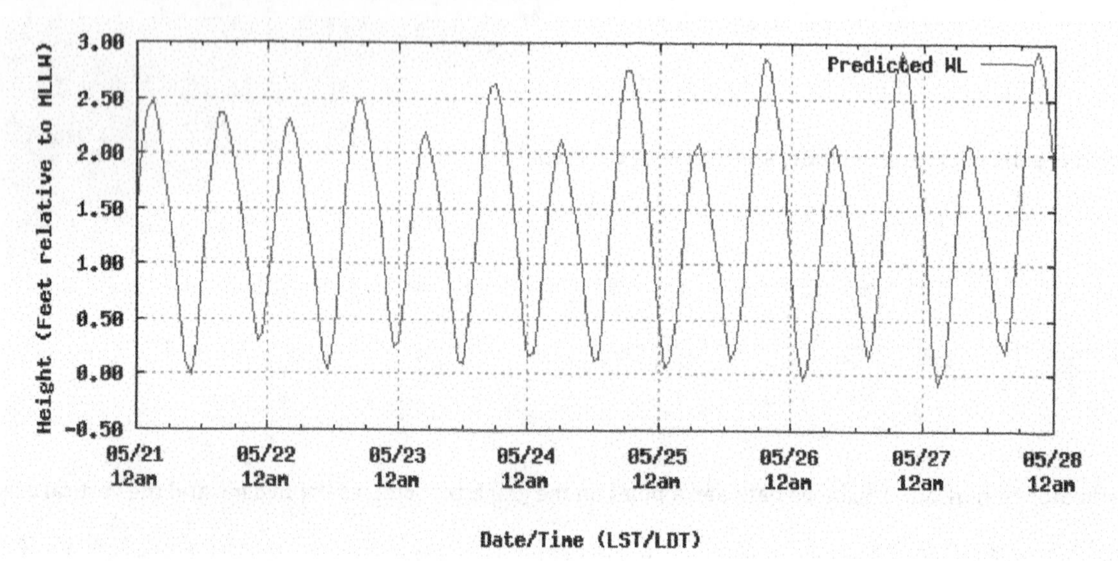

Source: http://tidesandcurrents.noaa.gov/

Example 2: Digital Sampling of Sound

When sound is recorded or transmitted electronically, the continuous waveform is sampled to convert it to a discrete digital sequence. If the sampling rate (represented by the horizontal scaling) or the resolution (represented by the vertical scaling) increases, the sound quality of the recording or transmission improves.

The data graphed below represent two different samples of a pure tone. One is sampled 8 times per unit of time at a 2-bit resolution (4 equal intervals in the vertical direction), and the other is sampled 16 times per unit of time at a 4-bit resolution (16 equal intervals in the vertical direction).

a. Which sample points would produce a better model of the actual sound wave? Explain your reasoning.

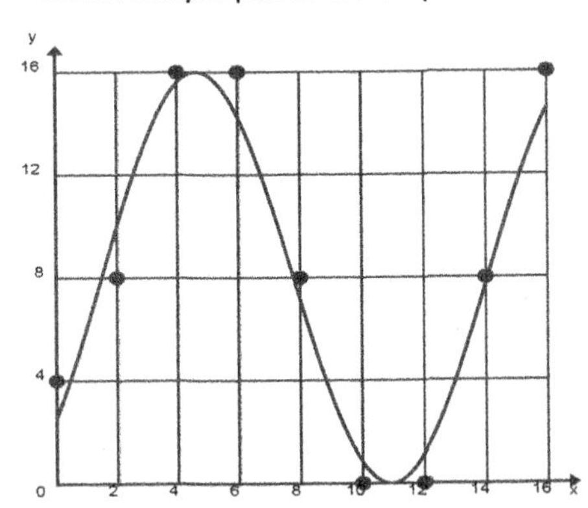

Figure 1: Sampled 8 times per unit of time, 2-bit resolution.

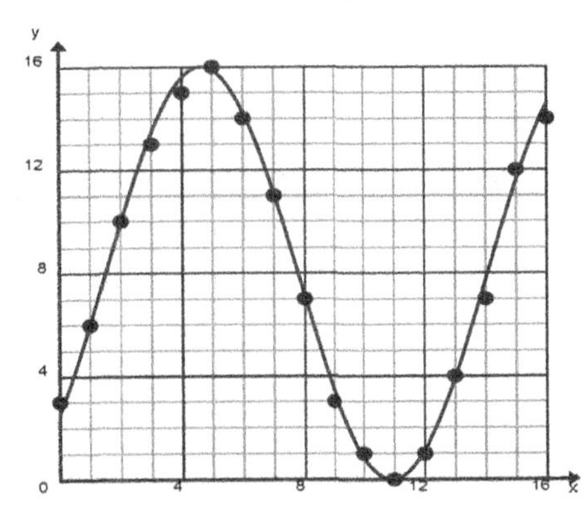

Figure 2: Sampled 16 times per unit of time, 4-bit resolution.

b. Find an equation for a function g that models the pure tone sampled in Figure 1.

c. Find an equation for a function h that models the pure tone sampled in Figure 2.

Exercises 2–6

Stock prices have historically increased over time, but they also vary in a cyclical fashion. Create a scatter plot of the data for the monthly stock price for a 15-month time period since January 1, 2003.

Months Since Jan. 1, 2003	Price at Close in dollars
0	20.24
1	19.42
2	18.25
3	19.13
4	19.20
5	20.91
6	20.86
7	20.04
8	20.30
9	20.54
10	21.94
11	21.87
12	21.51
13	20.65
14	19.84

2. Would a sinusoidal function be an appropriate model for these data? Explain your reasoning.

We can model the slight upward trend in these data with the linear function $f(x) = 19.5 + 0.13x$.

If we add a sinusoidal function to this linear function, we can model these data with an algebraic function that displays an upward trend but also varies in a cyclical fashion.

3. Find a sinusoidal function, g, that when added to the linear function, f, will model these data.

4. Let S be the stock price function that is the sum of the linear function listed above and the sinusoidal function in Exercise 3.

$$S(x) = \underline{\qquad\qquad\qquad\qquad}.$$

5. Add the graph of this function to your scatter plot. How well does it appear to fit the data?

6. Here is a graph of the same stock through December 2009.

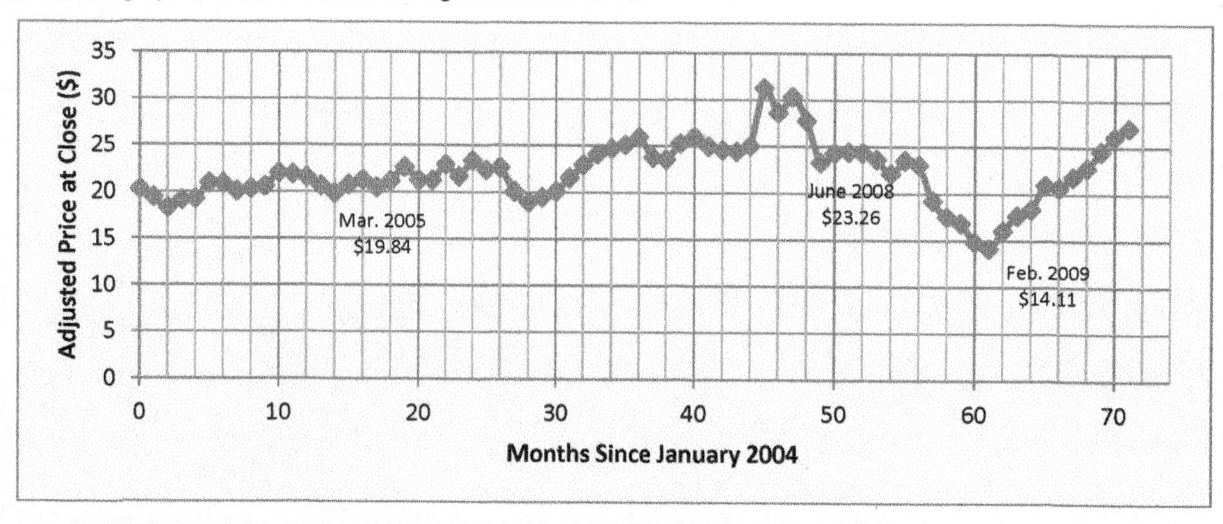

a. Will your model do a good job of predicting stock values past the year 2005?

b. What event occurred in 2008 to account for the sharp decline in the value of stocks?

c. What are the limitations of using any function to make predictions regarding the value of a stock at some point in the future?

Lesson Summary

Periodic data can be modeled with either a sine or a cosine function by extrapolating values of the parameters A, ω, h, and k from the data and defining a function $f(t) = A\sin(\omega(t-h)) + k$ or $g(t) = A\cos(\omega(t-h)) + k$, as appropriate.

Sine or cosine functions may not perfectly fit most data sets from actual measurements; therefore, there are often multiple functions used to model a data set.

If possible, plot the data together with the function that appears to fit the graph. If it is not a good fit, adjust the model and try again.

Problem Set

1. Find equations of both a sine function and a cosine function that could each represent the graph given below.

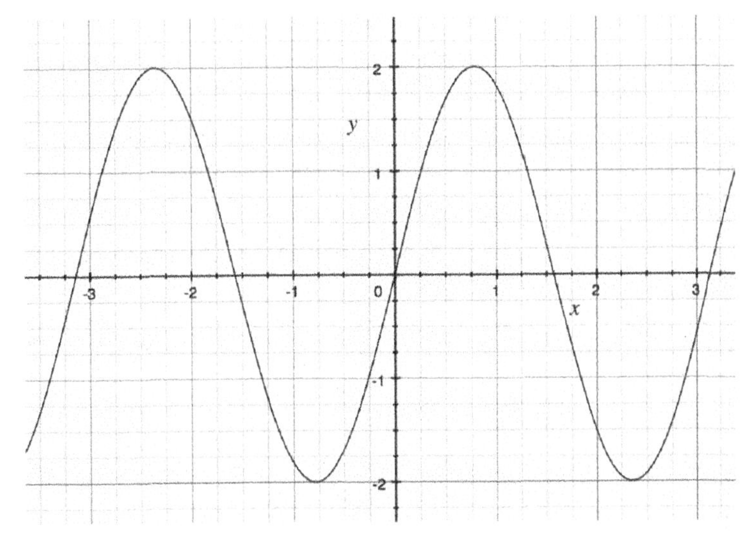

2. Find equations of both a sine function and a cosine function that could each represent the graph given below.

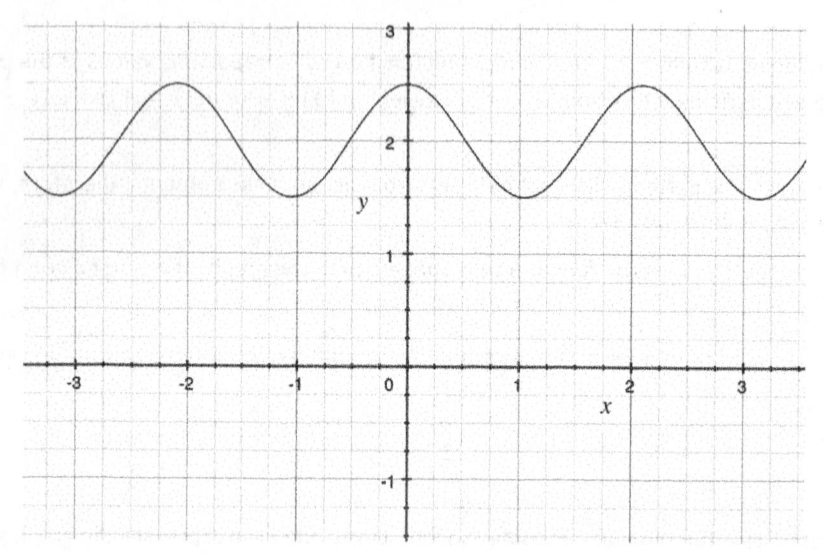

3. Rapidly vibrating objects send pressure waves through the air that are detected by our ears and then interpreted by our brains as sound. Our brains analyze the amplitude and frequency of these pressure waves.

   A speaker usually consists of a paper cone attached to an electromagnet. By sending an oscillating electric current through the electromagnet, the paper cone can be made to vibrate. By adjusting the current, the amplitude and frequency of vibrations can be controlled.

   The following graph shows the pressure intensity (I) as a function of time (x), in seconds, of the pressure waves emitted by a speaker set to produce a single pure tone.

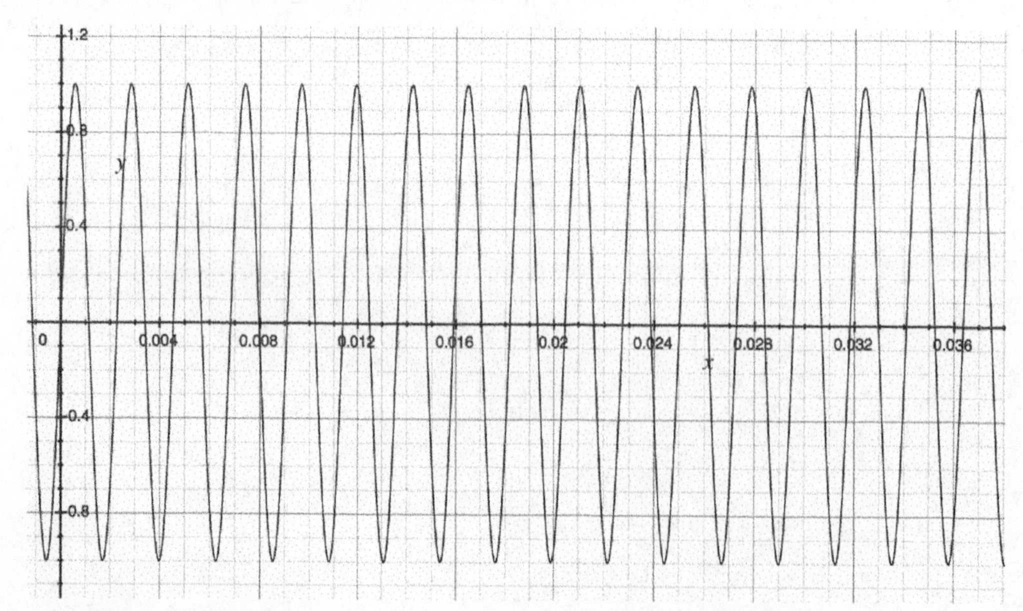

   a. Does it seem more natural to use a sine or a cosine function to fit to this graph?
   b. Find the equation of a trigonometric function that fits this graph.

4. Suppose that the following table represents the average monthly ambient air temperature, in degrees Fahrenheit, in some subterranean caverns in southeast Australia for each of the twelve months in a year. We wish to model these data with a trigonometric function. (Notice that the seasons are reversed in the Southern Hemisphere, so January is in summer, and July is in winter.)

Month	Jan.	Feb.	Mar.	Apr.	May	June	July	Aug.	Sept.	Oct.	Nov.	Dec.
°F	64.04	64.22	61.88	57.92	53.60	50.36	49.10	49.82	52.34	55.22	58.10	61.52

a. Does it seem reasonable to assume that these data, if extended beyond one year, should be roughly periodic?

b. What seems to be the amplitude of the data?

c. What seems to be the midline of the data (equation of the line through the middle of the graph)?

d. Would it be easier to use sine or cosine to model these data?

e. What is a reasonable approximation for the horizontal shift?

f. Write an equation for a function that could fit these data.

5. The table below provides data for the number of daylight hours as a function of day of the year, where day 1 represents January 1. Graph the data and determine if they could be represented by a sinusoidal function. If they can, determine the period, amplitude, and midline of the function, and find an equation for a function that models the data.

Day of Year	0	50	100	150	175	200	250	300	350
Hours	4.0	7.9	14.9	19.9	20.5	19.5	14.0	7.1	3.6

6. The function graphed below is $y = x^{\sin(x)}$. Blake says, "The function repeats on a fixed interval, so it must be a sinusoidal function." Respond to his argument.

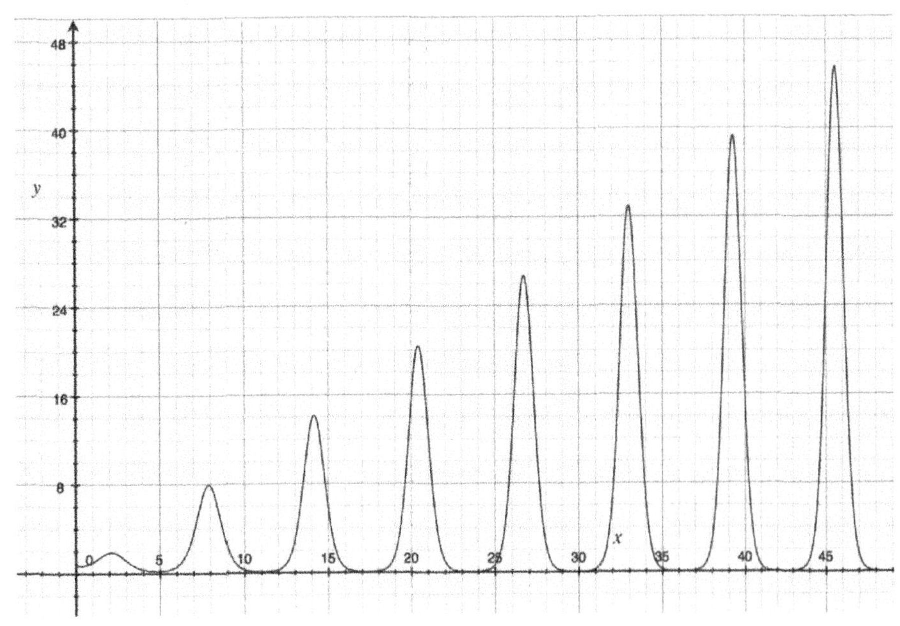

This page intentionally left blank

Lesson 14: Graphing the Tangent Function

Classwork

Exploratory Challenge/Exercises 1–5

1. Use your calculator to calculate each value of $\tan(x)$ to two decimal places in the table for your group.

Group 1 $\left(-\frac{\pi}{2}, \frac{\pi}{2}\right)$		Group 2 $\left(\frac{\pi}{2}, \frac{3\pi}{2}\right)$		Group 3 $\left(-\frac{3\pi}{2}, -\frac{\pi}{2}\right)$		Group 4 $\left(\frac{3\pi}{2}, \frac{5\pi}{2}\right)$	
x	$\tan(x)$	x	$\tan(x)$	x	$\tan(x)$	x	$\tan(x)$
$-\frac{11\pi}{24}$		$\frac{13\pi}{24}$		$-\frac{35\pi}{24}$		$\frac{37\pi}{24}$	
$-\frac{5\pi}{12}$		$\frac{7\pi}{12}$		$-\frac{17\pi}{12}$		$\frac{19\pi}{12}$	
$-\frac{4\pi}{12}$		$\frac{8\pi}{12}$		$-\frac{16\pi}{12}$		$\frac{20\pi}{12}$	
$-\frac{3\pi}{12}$		$\frac{9\pi}{12}$		$-\frac{15\pi}{12}$		$\frac{21\pi}{12}$	
$-\frac{2\pi}{12}$		$\frac{10\pi}{12}$		$-\frac{14\pi}{12}$		$\frac{22\pi}{12}$	
$-\frac{\pi}{12}$		$\frac{11\pi}{12}$		$-\frac{13\pi}{12}$		$\frac{23\pi}{12}$	
0		π		$-\pi$		2π	
$\frac{\pi}{12}$		$\frac{13\pi}{12}$		$-\frac{11\pi}{12}$		$\frac{25\pi}{12}$	
$\frac{2\pi}{12}$		$\frac{14\pi}{12}$		$-\frac{10\pi}{12}$		$\frac{26\pi}{12}$	
$\frac{3\pi}{12}$		$\frac{15\pi}{12}$		$-\frac{9\pi}{12}$		$\frac{27\pi}{12}$	
$\frac{4\pi}{12}$		$\frac{16\pi}{12}$		$-\frac{8\pi}{12}$		$\frac{28\pi}{12}$	
$\frac{5\pi}{12}$		$\frac{17\pi}{12}$		$-\frac{7\pi}{12}$		$\frac{29\pi}{12}$	
$\frac{11\pi}{24}$		$\frac{35\pi}{24}$		$-\frac{13\pi}{24}$		$\frac{59\pi}{24}$	

Group 5 $\left(-\frac{5\pi}{2}, -\frac{3\pi}{2}\right)$	
x	$\tan(x)$
$-\frac{59\pi}{24}$	
$-\frac{29\pi}{12}$	
$-\frac{28\pi}{12}$	
$-\frac{27\pi}{12}$	
$-\frac{26\pi}{12}$	
$-\frac{25\pi}{12}$	
-2π	
$-\frac{23\pi}{12}$	
$-\frac{22\pi}{12}$	
$-\frac{21\pi}{12}$	
$-\frac{20\pi}{12}$	
$-\frac{19\pi}{12}$	
$-\frac{37\pi}{24}$	

Group 6 $\left(\frac{5\pi}{2}, \frac{7\pi}{2}\right)$	
x	$\tan(x)$
$\frac{61\pi}{24}$	
$\frac{31\pi}{12}$	
$\frac{32\pi}{12}$	
$\frac{33\pi}{12}$	
$\frac{34\pi}{12}$	
$\frac{35\pi}{12}$	
3π	
$\frac{37\pi}{12}$	
$\frac{38\pi}{12}$	
$\frac{39\pi}{12}$	
$\frac{40\pi}{12}$	
$\frac{41\pi}{12}$	
$\frac{83\pi}{24}$	

Group 7 $\left(-\frac{7\pi}{2}, -\frac{5\pi}{2}\right)$	
x	$\tan(x)$
$-\frac{83\pi}{24}$	
$-\frac{41\pi}{12}$	
$-\frac{40\pi}{12}$	
$-\frac{39\pi}{12}$	
$-\frac{38\pi}{12}$	
$-\frac{37\pi}{12}$	
-3π	
$-\frac{35\pi}{12}$	
$-\frac{34\pi}{12}$	
$-\frac{33\pi}{12}$	
$-\frac{32\pi}{12}$	
$-\frac{31\pi}{12}$	
$-\frac{61\pi}{24}$	

Group 8 $\left(\frac{7\pi}{2}, \frac{9\pi}{2}\right)$	
x	$\tan(x)$
$\frac{37\pi}{24}$	
$\frac{43\pi}{12}$	
$\frac{44\pi}{12}$	
$\frac{45\pi}{12}$	
$\frac{46\pi}{12}$	
$\frac{47\pi}{12}$	
4π	
$\frac{49\pi}{12}$	
$\frac{50\pi}{12}$	
$\frac{51\pi}{12}$	
$\frac{52\pi}{12}$	
$\frac{53\pi}{12}$	
$\frac{107\pi}{24}$	

2. The tick marks on the axes provided are spaced in increments of $\frac{\pi}{12}$. Mark the horizontal axis by writing the number of the left endpoint of your interval at the leftmost tick mark, the multiple of π that is in the middle of your interval at the point where the axes cross, and the number at the right endpoint of your interval at the rightmost tick mark. Fill in the remaining values at increments of $\frac{\pi}{12}$.

3. On your plot, sketch the graph of $y = \tan(x)$ on your specified interval by plotting the points in the table and connecting the points with a smooth curve. Draw the graph with a bold marker.

4. What happens to the graph near the edges of your interval? Why does this happen?

5. When you are finished, affix your graph to the board in the appropriate place, matching endpoints of intervals.

Exploratory Challenge 2/Exercises 6–16

For each exercise below, let $m = \tan(\theta)$ be the slope of the terminal ray in the definition of the tangent function, and let $P = (x_0, y_0)$ be the intersection of the terminal ray with the unit circle after being rotated by θ radians for $0 < \theta < \frac{\pi}{2}$. We know that the tangent of θ is the slope m of $\overleftrightarrow{OP}$.

6. Let Q be the intersection of the terminal ray with the unit circle after being rotated by $\theta + \pi$ radians.

   a. What is the slope of $\overleftrightarrow{OQ}$?

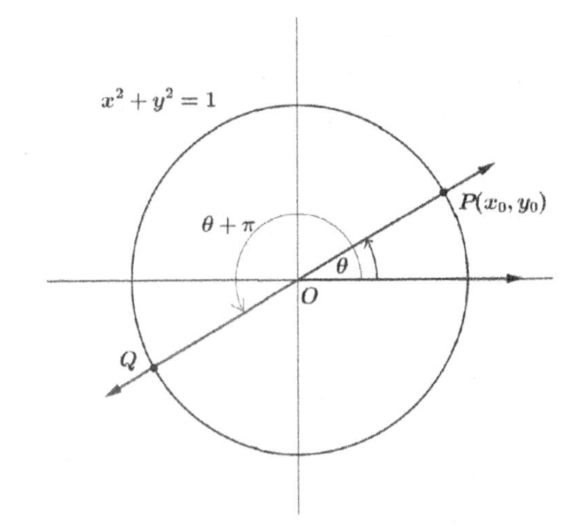

   b. Find an expression for $\tan(\theta + \pi)$ in terms of m.

   c. Find an expression for $\tan(\theta + \pi)$ in terms of $\tan(\theta)$.

   d. How can the expression in part (c) be seen in the graph of the tangent function?

7. Let Q be the intersection of the terminal ray with the unit circle after being rotated by $-\theta$ radians.

   a. What is the slope of $\overleftrightarrow{OQ}$?

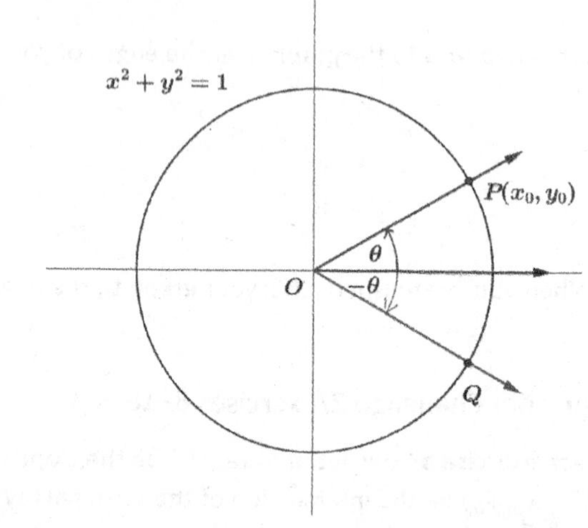

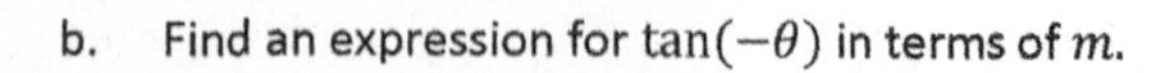

   b. Find an expression for $\tan(-\theta)$ in terms of m.

   c. Find an expression for $\tan(-\theta)$ in terms of $\tan(\theta)$.

   d. How can the expression in part (c) be seen in the graph of the tangent function?

8. Is the tangent function an even function, an odd function, or neither? How can you tell your answer is correct from the graph of the tangent function?

9. Let Q be the intersection of the terminal ray with the unit circle after being rotated by $\pi - \theta$ radians.

   a. What is the slope of $\overleftrightarrow{OQ}$?

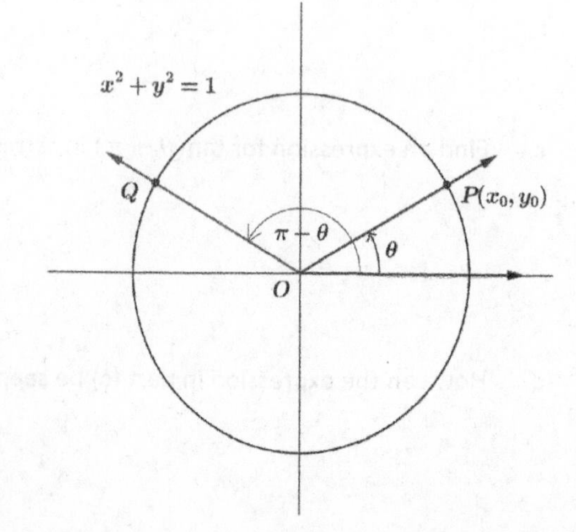

   b. Find an expression for $\tan(\pi - \theta)$ in terms of m.

   c. Find an expression for $\tan(\pi - \theta)$ in terms of $\tan(\theta)$.

10. Let Q be the intersection of the terminal ray with the unit circle after being rotated by $\frac{\pi}{2} + \theta$ radians.

    a. What is the slope of $\overleftrightarrow{OQ}$?

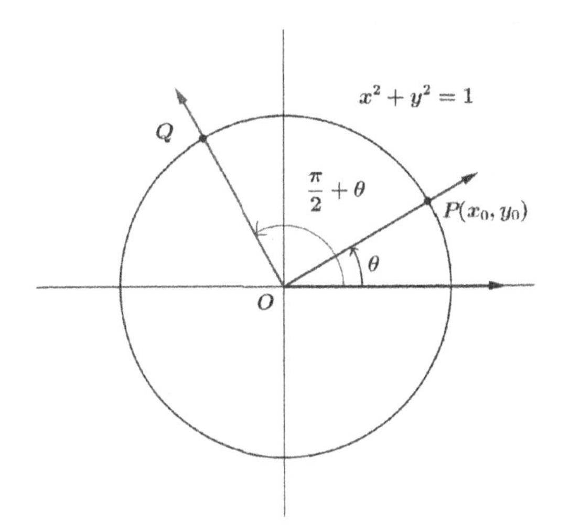

    b. Find an expression for $\tan\left(\frac{\pi}{2} + \theta\right)$ in terms of m.

    c. Find an expression for $\tan\left(\frac{\pi}{2} + \theta\right)$ first in terms of $\tan(\theta)$ and then in terms of $\cot(\theta)$.

11. Let Q be the intersection of the terminal ray with the unit circle after being rotated by $\frac{\pi}{2} - \theta$ radians.

    a. What is the slope of $\overleftrightarrow{OQ}$?

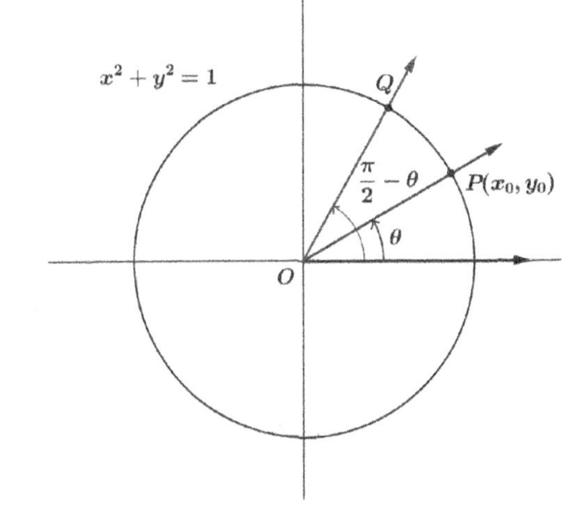

    b. Find an expression for $\tan\left(\frac{\pi}{2} - \theta\right)$ in terms of m.

    c. Find an expression for $\tan\left(\frac{\pi}{2} - \theta\right)$ in terms of $\tan(\theta)$ or other trigonometric functions.

12. Summarize your results from Exercises 6, 7, 9, 10, and 11.

13. We have only demonstrated that the identities in Exercise 12 are valid for $0 < \theta < \frac{\pi}{2}$ because we only used rotations that left point P in the first quadrant. Argue that $\tan\left(-\frac{2\pi}{3}\right) = -\tan\left(\frac{2\pi}{3}\right)$. Then, using similar logic, we could argue that all of the above identities extend to any value of θ for which the tangent (and cotangent for the last two) is defined.

14. For which values of θ are the identities in Exercise 12 valid?

15. Derive an identity for $\tan(2\pi + \theta)$ from the graph.

16. Use the identities you summarized in Exercise 12 to show $\tan(2\pi - \theta) = -\tan(\theta)$ where $\theta \neq \frac{\pi}{2} + k\pi$, for all integers k.

Lesson Summary

The tangent function $\tan(x) = \frac{\sin(x)}{\cos(x)}$ is periodic with period π. The following identities have been established.

- $\tan(x + \pi) = \tan(x)$ for all $x \neq \frac{\pi}{2} + k\pi$, for all integers k.
- $\tan(-x) = -\tan(x)$ for all $x \neq \frac{\pi}{2} + k\pi$, for all integers k.
- $\tan(\pi - x) = -\tan(x)$ for all $x \neq \frac{\pi}{2} + k\pi$, for all integers k.
- $\tan\left(\frac{\pi}{2} + x\right) = -\cot(x)$ for all $x \neq k\pi$, for all integers k.
- $\tan\left(\frac{\pi}{2} - x\right) = \cot(x)$ for all $x \neq k\pi$, for all integers k.
- $\tan(2\pi + x) = \tan(x)$ for all $x \neq \frac{\pi}{2} + k\pi$, for all integers k.
- $\tan(2\pi - x) = -\tan(x)$ for all $x \neq \frac{\pi}{2} + k\pi$, for all integers k.

Problem Set

1. Recall that the cotangent function is defined by $\cot(x) = \frac{\cos(x)}{\sin(x)} = \frac{1}{\tan(x)}$, where $\sin(x) \neq 0$.

   a. What is the domain of the cotangent function? Explain how you know.

   b. What is the period of the cotangent function? Explain how you know.

   c. Use a calculator to complete the table of values of the cotangent function on the interval $(0, \pi)$ to two decimal places.

x	$\cot(x)$
$\frac{\pi}{24}$	
$\frac{\pi}{12}$	
$\frac{2\pi}{12}$	
$\frac{3\pi}{12}$	

x	$\cot(x)$
$\frac{4\pi}{12}$	
$\frac{5\pi}{12}$	
$\frac{\pi}{2}$	

x	$\cot(x)$
$\frac{7\pi}{12}$	
$\frac{8\pi}{12}$	
$\frac{9\pi}{12}$	

x	$\cot(x)$
$\frac{10\pi}{12}$	
$\frac{11\pi}{12}$	
$\frac{23\pi}{24}$	

   d. Plot your data from part (c), and sketch a graph of $y = \cot(x)$ on $(0, \pi)$.

   e. Sketch a graph of $y = \cot(x)$ on $(-2\pi, 2\pi)$ without plotting points.

   f. Discuss the similarities and differences between the graphs of the tangent and cotangent functions.

   g. Find all x-values where $\tan(x) = \cot(x)$ on the interval $(0, 2\pi)$.

2. Each set of axes below shows the graph of $f(x) = \tan(x)$. Use what you know about function transformations to sketch a graph of $y = g(x)$ for each function g on the interval $(0, 2\pi)$.

a. $g(x) = 2\tan(x)$

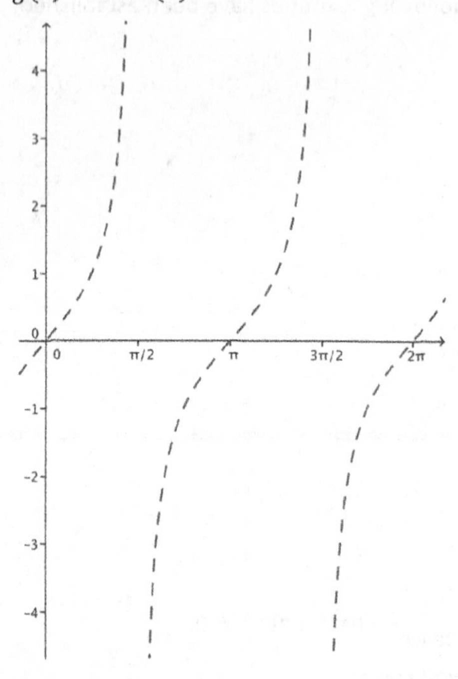

b. $g(x) = \frac{1}{3}\tan(x)$

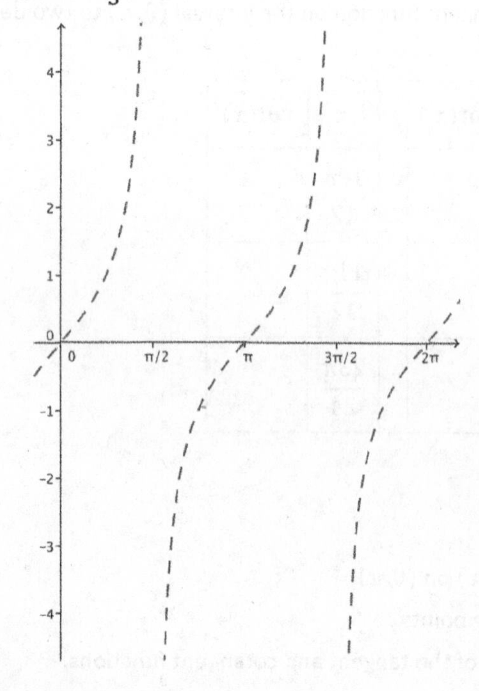

c. $g(x) = -2\tan(x)$

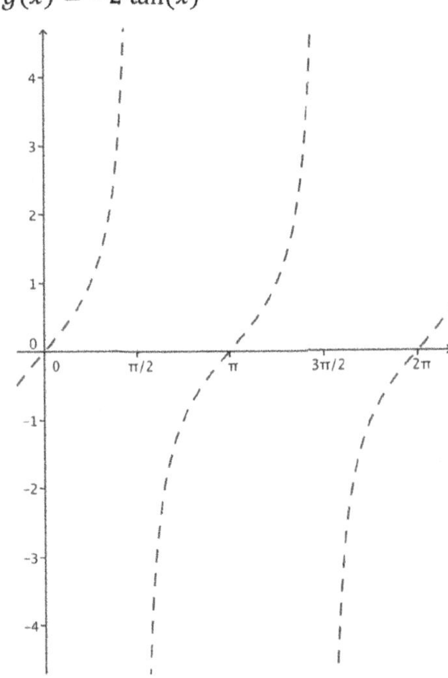

d. How does changing the parameter A affect the graph of $g(x) = A\tan(x)$?

3. Each set of axes below shows the graph of $f(x) = \tan(x)$. Use what you know about function transformations to sketch a graph of $y = g(x)$ for each function g on the interval $(0, 2\pi)$.

a. $g(x) = \tan\left(x - \frac{\pi}{2}\right)$

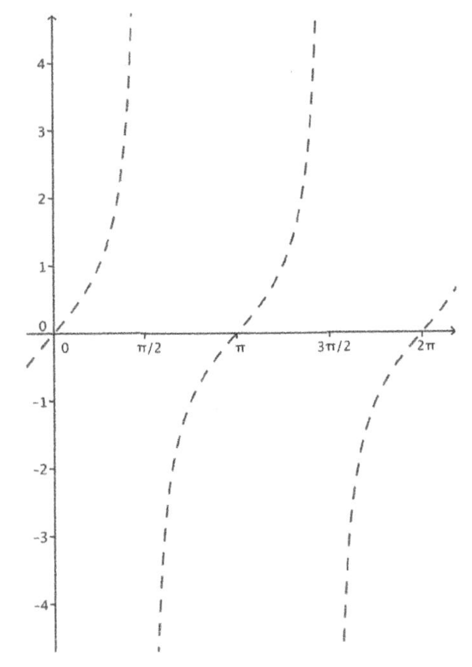

b. $g(x) = \tan\left(x - \frac{\pi}{6}\right)$

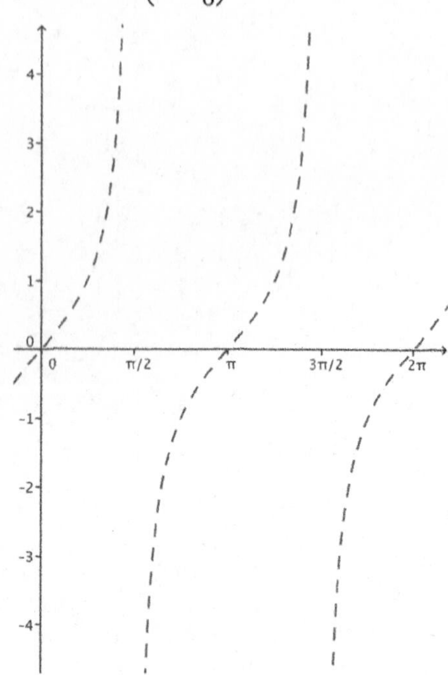

c. $g(x) = \tan\left(x + \frac{\pi}{4}\right)$

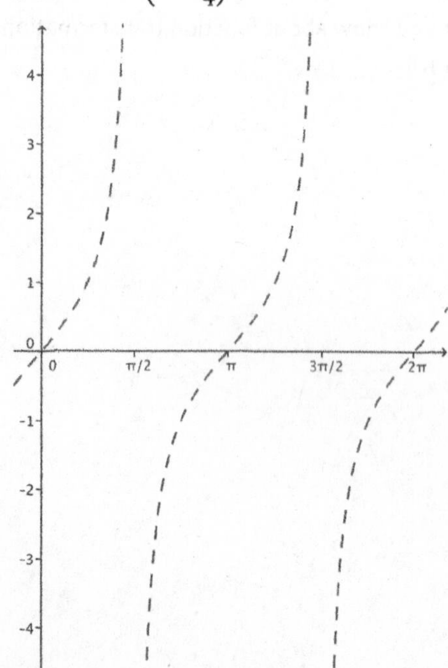

d. **How does changing the parameter h affect the graph of $g(x) = \tan(x - h)$?**

4. Each set of axes below shows the graph of $f(x) = \tan(x)$. Use what you know about function transformations to sketch a graph of $y = g(x)$ for each function g on the interval $(0, 2\pi)$.

   a. $g(x) = \tan(x) + 1$

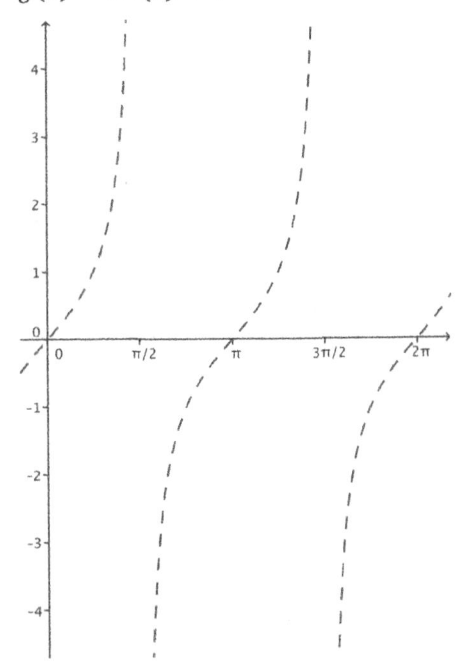

   b. $g(x) = \tan(x) + 3$

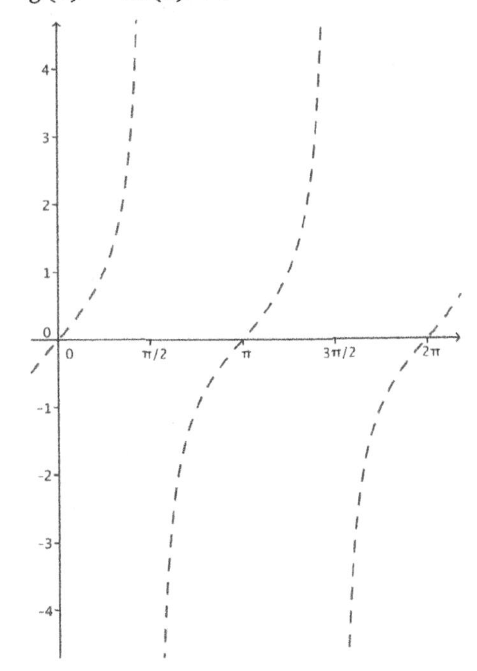

c. $g(x) = \tan(x) - 2$

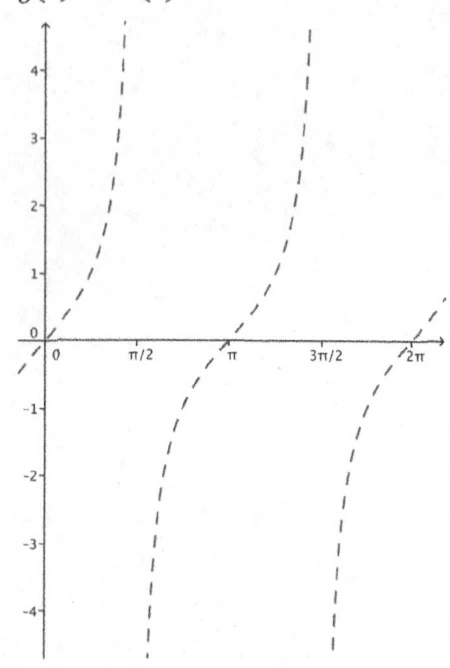

d. How does changing the parameter k affect the graph of $g(x) = \tan(x) + k$?

5. Each set of axes below shows the graph of $f(x) = \tan(x)$. Use what you know about function transformations to sketch a graph of $y = g(x)$ for each function g on the interval $(0, 2\pi)$.

a. $g(x) = \tan(3x)$

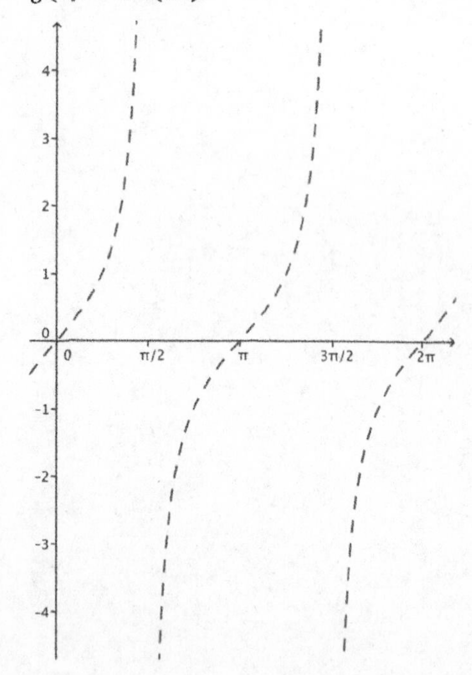

b. $g(x) = \tan\left(\frac{x}{2}\right)$

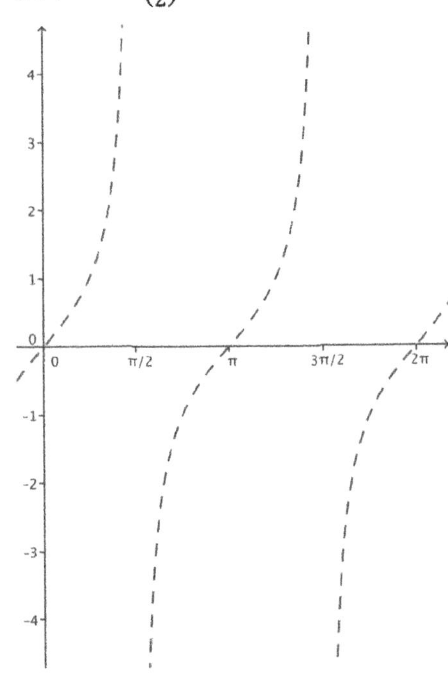

c. $g(x) = \tan(-3x)$

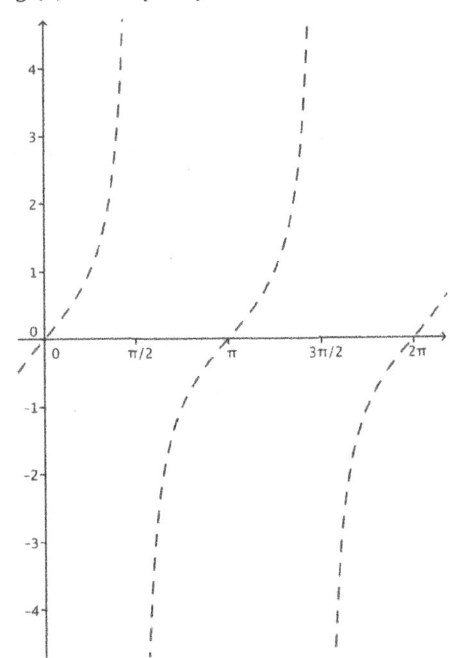

d. How does changing the parameter ω affect the graph of $g(x) = \tan(\omega x)$?

6. Use your knowledge of function transformation and the graph of $y = \tan(x)$ to sketch graphs of the following transformations of the tangent function.
   a. $y = \tan(2x)$
   b. $y = \tan\left(2\left(x - \frac{\pi}{4}\right)\right)$
   c. $y = \tan\left(2\left(x - \frac{\pi}{4}\right)\right) + 1.5$

7. Find parameters A, ω, h, and k so that the graphs of $f(x) = A\tan(\omega(x - h)) + k$ and $g(x) = \cot(x)$ are the same.

Lesson 15: What Is a Trigonometric Identity?

Classwork

Exercises 1–3

1. Recall the Pythagorean identity $\sin^2(\theta) + \cos^2(\theta) = 1$, where θ is any real number.

   a. Find $\sin(x)$, given $\cos(x) = \frac{3}{5}$, for $-\frac{\pi}{2} < x < 0$.

   b. Find $\tan(y)$, given $\cos(y) = -\frac{5}{13}$, for $\frac{\pi}{2} < y < \pi$.

   c. Write $\tan(z)$ in terms of $\cos(z)$, for $\pi < z < \frac{3\pi}{2}$.

2. Use the Pythagorean identity to do the following:

   a. Rewrite the expression $\cos(\theta)\sin^2(\theta) - \cos(\theta)$ in terms of a single trigonometric function. State the resulting identity.

   b. Rewrite the expression $(1 - \cos^2(\theta))\csc(\theta)$ in terms of a single trigonometric function. State the resulting identity.

   c. Find all solutions to the equation $2\sin^2(\theta) = 2 + \cos(\theta)$ in the interval $(0, 2\pi)$. Draw a unit circle that shows the solutions.

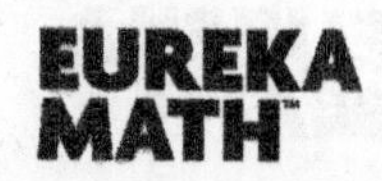

3. Which of the following statements are identities? If a statement is an identity, specify the values of x where the equation holds.

   a. $\sin(x + 2\pi) = \sin(x)$ where the functions on both sides are defined.

   b. $\sec(x) = 1$ where the functions on both sides are defined.

   c. $\sin(-x) = \sin(x)$ where the functions on both sides are defined.

   d. $1 + \tan^2(x) = \sec^2(x)$ where the functions on both sides are defined.

   e. $\sin\left(\frac{\pi}{2} - x\right) = \cos(x)$ where the functions on both sides are defined.

   f. $\sin^2(x) = \tan^2(x)$ for all real x.

Lesson Summary

The Pythagorean identity: $\sin^2(\theta) + \cos^2(\theta) = 1$ for all real numbers θ.

Problem Set

1. Which of the following statements are trigonometric identities? Graph the functions on each side of the equation.
   a. $\tan(x) = \frac{\sin(x)}{\cos(x)}$ where the functions on both sides are defined.
   b. $\cos^2(x) = 1 + \sin(x)$ where the functions on both sides are defined.
   c. $\cos\left(\frac{\pi}{2} - x\right) = \sin(x)$ where the functions on both sides are defined.

2. Determine the domain of the following trigonometric identities:
   a. $\cot(x) = \frac{\cos(x)}{\sin(x)}$ where the functions on both sides are defined.
   b. $\cos(-u) = \cos(u)$ where the functions on both sides are defined.
   c. $\sec(y) = \frac{1}{\cos(y)}$ where the functions on both sides are defined.

3. Rewrite $\sin(x)\cos^2(x) - \sin(x)$ as an expression containing a single term.

4. Suppose $0 < \theta < \frac{\pi}{2}$ and $\sin(\theta) = \frac{1}{\sqrt{3}}$. What is the value of $\cos(\theta)$?

5. If $\cos(\theta) = -\frac{1}{\sqrt{5}}$, what are possible values of $\sin(\theta)$?

6. Use the Pythagorean identity $\sin^2(\theta) + \cos^2(\theta) = 1$, where θ is any real number, to find the following:
   a. $\cos(\theta)$, given $\sin(\theta) = \frac{5}{13}$, for $\frac{\pi}{2} < \theta < \pi$.
   b. $\tan(x)$, given $\cos(x) = -\frac{1}{\sqrt{2}}$, for $\pi < x < \frac{3\pi}{2}$.

7. The three identities below are all called Pythagorean identities. The second and third follow from the first, as you saw in Example 1 and the Exit Ticket.
   a. For which values of θ are each of these identities defined?
      i. $\sin^2(\theta) + \cos^2(\theta) = 1$, where the functions on both sides are defined.
      ii. $\tan^2(\theta) + 1 = \sec^2(\theta)$, where the functions on both sides are defined.
      iii. $1 + \cot^2(\theta) = \csc^2(\theta)$, where the functions on both sides are defined.
   b. For which of the three identities is 0 in the domain of validity?
   c. For which of the three identities is $\frac{\pi}{2}$ in the domain of validity?
   d. For which of the three identities is $-\frac{\pi}{4}$ in the domain of validity?

This page intentionally left blank

Lesson 16: Proving Trigonometric Identities

Classwork

Opening Exercise

Which of these statements is a trigonometric identity? Provide evidence to support your claim.

Statement 1: $\sin^2(\theta) = 1 - \cos^2(\theta)$ for θ any real number.

Statement 2: $1 - \cos(\theta) = 1 - \cos(\theta)$ for θ any real number.

Statement 3: $1 - \cos(\theta) = 1 + \cos(\theta)$ for θ any real number.

Using Statements 1 and 2, create a third identity, Statement 4, whose left side is $\frac{\sin^2(\theta)}{1-\cos(\theta)}$.

For which values of θ is this statement valid?

Discuss in pairs what it might mean to "prove" an identity. What might it take to prove, for example, that the following statement is an identity?

$$\frac{\sin^2(\theta)}{1-\cos(\theta)} = 1 + \cos(\theta) \text{ where } \theta \neq 2\pi k, \text{ for all integers } k.$$

To prove an identity, you have to use logical steps to show that one side of the equation in the identity can be transformed into the other side of the equation using already established identities such as the Pythagorean identity or the properties of operation (commutative, associative, and distributive properties). It is not correct to start with what you want to prove and work on both sides of the equation at the same time, as the following exercise shows.

Exercise 1

1. Use a calculator to graph the functions $f(x) = \sin(x) + \cos(x)$ and $g(x) = -\sqrt{1 + 2\sin(x)\cos(x)}$ to determine whether $\sin(\theta) + \cos(\theta) = -\sqrt{1 + 2\sin(\theta)\cos(\theta)}$ for all θ for which both functions are defined is a valid identity. You should see from the graphs that the functions are not equivalent.

   Suppose that Charles did not think to graph the equations to see if the given statement was a valid identity, so he set about proving the identity using algebra and a previous identity. His argument is shown below.

   First, [1] $\sin(\theta) + \cos(\theta) = -\sqrt{1 + 2\sin(\theta)\cos(\theta)}$ for θ any real number.

   Now, using the multiplication property of equality, square both sides, which gives

   [2] $\sin^2(\theta) + 2\sin(\theta)\cos(\theta) + \cos^2(\theta) = 1 + 2\sin(\theta)\cos(\theta)$ for θ any real number.

   Using the subtraction property of equality, subtract $2\sin(\theta)\cos(\theta)$ from each side, which gives

   [3] $\sin^2(\theta) + \cos^2(\theta) = 1$ for θ any real number.

   Statement [3] is the Pythagorean identity. So, replace $\sin^2(\theta) + \cos^2(\theta)$ by 1 to get

   [4] $1 = 1$, which is definitely true.

   Therefore, the original statement must be true.

   Does this mean that Charles has proven that Statement [1] is an identity? Discuss with your group whether it is a valid proof. If you decide it is not a valid proof, then discuss with your group how and where his argument went wrong.

Example 1: Two Proofs of Our New Identity

Work through these two different ways to approach proving the identity $\frac{\sin^2(\theta)}{1-\cos(\theta)} = 1 + \cos(\theta)$ where $\theta \neq 2\pi k$, for integers k. The proofs make use of some of the following properties of equality and real numbers. Here a, b, and c stand for arbitrary real numbers.

Reflexive property of equality	$a = a$
Symmetric property of equality	If $a = b$, then $b = a$.
Transitive property of equality	If $a = b$ and $b = c$, then $a = c$.
Addition property of equality	If $a = b$, then $a + c = b + c$.
Subtraction property of equality	If $a = b$, then $a - c = b - c$.
Multiplication property of equality	If $a = b$, then $a \cdot c = b \cdot c$.
Division property of equality	If $a = b$ and $c \neq 0$, then $a \div c = b \div c$.
Substitution property of equality	If $a = b$, then b may be substituted for a in any expression containing a.
Associative properties	$(a + b) + c = a + (b + c)$ and $a(bc) = (ab)c$.
Commutative properties	$a + b = b + a$ and $ab = ba$.
Distributive property	$a(b + c) = ab + ac$ and $(a + b)c = ac + bc$.

Fill in the missing parts of the proofs outlined in the tables below. Then, write a proof of the resulting identity.

a. We start with the Pythagorean identity. When we divide both sides by the same expression, $1 - \cos(\theta)$, we introduce potential division by zero when $\cos(\theta) = 1$. This will change the set of values of θ for which the identity is valid.

PROOF:

Step	Left Side of Equation		Equivalent Right Side	Domain	Reason
1	$\sin^2(\theta) + \cos^2(\theta)$	$=$	1	θ any real number	Pythagorean identity
2	$\sin^2(\theta)$	$=$	$1 - \cos^2(\theta)$	θ any real number	
3		$=$	$(1 - \cos(\theta))(1 + \cos(\theta))$	θ any real number	
4		$=$	$\frac{(1 - \cos(\theta))(1 + \cos(\theta))}{1 - \cos(\theta)}$		
5	$\frac{\sin^2(\theta)}{1 - \cos(\theta)}$	$=$		$\theta \neq 2\pi k$, for all integers k	Substitution property of equality using $\frac{1-\cos(\theta)}{1-\cos(\theta)} = 1$

b. Or, we can start with the more complicated side of the identity we want to prove and use algebra and prior trigonometric definitions and identities to transform it to the other side. In this case, the more complicated expression is $\frac{\sin^2(\theta)}{1-\cos(\theta)}$.

PROOF:

Step	Left Side of Equation		Equivalent Right Side	Domain	Reason
1	$\frac{\sin^2(\theta)}{1-\cos(\theta)}$	$=$	$\frac{1-\cos^2(\theta)}{1-\cos(\theta)}$	$\theta \neq 2\pi k$, for all integers k	Substitution property of equality using $\sin^2(\theta) = 1 - \cos^2(\theta)$
2		$=$	$\frac{(1-\cos(\theta))(1+\cos(\theta))}{1-\cos(\theta)}$		Distributive property
3	$\frac{\sin^2(\theta)}{1-\cos(\theta)}$	$=$	$1+\cos(\theta)$		

Exercises 2–3

Prove that the following are trigonometric identities, beginning with the side of the equation that seems to be more complicated and starting the proof by restricting x to values where the identity is valid. Make sure that the complete identity statement is included at the end of the proof.

2. $\tan(x) = \frac{\sec(x)}{\csc(x)}$ for real numbers $x \neq \frac{\pi}{2} + \pi k$, for all integers k.

3. $\cot(x) + \tan(x) = \sec(x)\csc(x)$ for all real numbers $x \neq \frac{\pi}{2}n$ for integer n.

Problem Set

1. Does $\sin(x+y)$ equal $\sin(x)+\sin(y)$ for all real numbers x and y?
   a. Find each of the following: $\sin\left(\frac{\pi}{2}\right), \sin\left(\frac{\pi}{4}\right), \sin\left(\frac{3\pi}{4}\right)$.
   b. Are $\sin\left(\frac{\pi}{2}+\frac{\pi}{4}\right)$ and $\sin\left(\frac{\pi}{2}\right)+\sin\left(\frac{\pi}{4}\right)$ equal?
   c. Are there any values of x and y for which $\sin(x+y)=\sin(x)+\sin(y)$?

2. Use $\tan(x)=\frac{\sin(x)}{\cos(x)}$ and identities involving the sine and cosine functions to establish the following identities for the tangent function. Identify the values of x where the equation is an identity.
   a. $\tan(\pi-x)=\tan(x)$
   b. $\tan(x+\pi)=\tan(x)$
   c. $\tan(2\pi-x)=-\tan(x)$
   d. $\tan(-x)=-\tan(x)$

3. Rewrite each of the following expressions as a single term. Identify the values of x for which the original expression and your expression are equal:
   a. $\cot(x)\sec(x)\sin(x)$
   b. $\left(\frac{1}{1-\sin(x)}\right)\left(\frac{1}{1+\sin(x)}\right)$
   c. $\frac{1}{\cos^2(x)}-\frac{1}{\cot^2(x)}$
   d. $\frac{(\tan(x)-\sin(x))(1+\cos(x))}{\sin^3(x)}$

4. Prove that for any two real numbers a and b,
$$\sin^2(a)-\sin^2(b)+\cos^2(a)\sin^2(b)-\sin^2(a)\cos^2(b)=0.$$

5. Prove that the following statements are identities for all values of θ for which both sides are defined, and describe that set.
   a. $\cot(\theta)\sec(\theta)=\csc(\theta)$
   b. $(\csc(\theta)+\cot(\theta))(1-\cos(\theta))=\sin(\theta)$
   c. $\tan^2(\theta)-\sin^2(\theta)=\tan^2(\theta)\sin^2(\theta)$
   d. $\frac{4+\tan^2(x)-\sec^2(x)}{\csc^2(x)}=3\sin^2(x)$
   e. $\frac{(1+\sin(\theta))^2+\cos^2(\theta)}{1+\sin(\theta)}=2$

6. Prove that the value of the following expression does not depend on the value of y:
$$\cot(y)\frac{\tan(x)+\tan(y)}{\cot(x)+\cot(y)}.$$

Lesson 17: Trigonometric Identity Proofs

Classwork

Opening Exercise

We have seen that $\sin(\alpha+\beta) \neq \sin(\alpha)+\sin(\beta)$. So, what is $\sin(\alpha+\beta)$? Begin by completing the following table:

α	β	$\sin(\alpha)$	$\sin(\beta)$	$\sin(\alpha+\beta)$	$\sin(\alpha)\cos(\beta)$	$\sin(\alpha)\sin(\beta)$	$\cos(\alpha)\cos(\beta)$	$\cos(\alpha)\sin(\beta)$
$\frac{\pi}{6}$	$\frac{\pi}{6}$	$\frac{1}{2}$	$\frac{1}{2}$	$\frac{\sqrt{3}}{2}$	$\frac{\sqrt{3}}{4}$	$\frac{1}{2}$		
$\frac{\pi}{6}$	$\frac{\pi}{3}$	$\frac{1}{2}$	$\frac{\sqrt{3}}{2}$	1			$\frac{\sqrt{3}}{4}$	
$\frac{\pi}{4}$	$\frac{\pi}{6}$	$\frac{\sqrt{2}}{2}$	$\frac{1}{2}$	$\frac{\sqrt{2}+\sqrt{6}}{4}$	$\frac{\sqrt{6}}{4}$			
$\frac{\pi}{4}$	$\frac{\pi}{4}$	$\frac{\sqrt{2}}{2}$	$\frac{\sqrt{2}}{2}$	1		$\frac{1}{2}$	$\frac{1}{2}$	
$\frac{\pi}{3}$	$\frac{\pi}{3}$	$\frac{\sqrt{3}}{2}$	$\frac{\sqrt{3}}{2}$	$\frac{\sqrt{3}}{2}$				$\frac{\sqrt{3}}{4}$
$\frac{\pi}{3}$	$\frac{\pi}{4}$	$\frac{\sqrt{3}}{2}$	$\frac{\sqrt{2}}{2}$	$\frac{\sqrt{2}+\sqrt{6}}{4}$		$\frac{\sqrt{6}}{4}$	$\frac{\sqrt{2}}{4}$	

Use the following table to formulate a conjecture for $\cos(\alpha + \beta)$:

α	β	$\cos(\alpha)$	$\cos(\beta)$	$\cos(\alpha+\beta)$	$\sin(\alpha)\cos(\beta)$	$\sin(\alpha)\sin(\beta)$	$\cos(\alpha)\cos(\beta)$	$\cos(\alpha)\sin(\beta)$
$\frac{\pi}{6}$	$\frac{\pi}{6}$	$\frac{1}{2}$	$\frac{1}{2}$	$\frac{1}{2}$	$\frac{\sqrt{3}}{4}$	$\frac{1}{2}$	$\frac{3}{4}$	$\frac{\sqrt{3}}{4}$
$\frac{\pi}{6}$	$\frac{\pi}{3}$	$\frac{\sqrt{3}}{2}$	$\frac{1}{2}$	0	$\frac{1}{4}$	$\frac{\sqrt{3}}{4}$	$\frac{\sqrt{3}}{4}$	$\frac{3}{4}$
$\frac{\pi}{4}$	$\frac{\pi}{6}$	$\frac{\sqrt{2}}{2}$	$\frac{\sqrt{3}}{2}$	$\frac{\sqrt{6}-\sqrt{2}}{4}$	$\frac{\sqrt{6}}{4}$	$\frac{\sqrt{2}}{4}$	$\frac{\sqrt{6}}{4}$	$\frac{\sqrt{2}}{4}$
$\frac{\pi}{4}$	$\frac{\pi}{4}$	$\frac{\sqrt{2}}{2}$	$\frac{\sqrt{2}}{2}$	0	$\frac{1}{2}$	$\frac{1}{2}$	$\frac{1}{2}$	$\frac{1}{2}$
$\frac{\pi}{3}$	$\frac{\pi}{3}$	$\frac{1}{2}$	$\frac{1}{2}$	$-\frac{1}{2}$	$\frac{\sqrt{3}}{4}$	$\frac{3}{4}$	$\frac{1}{4}$	$\frac{\sqrt{3}}{4}$
$\frac{\pi}{3}$	$\frac{\pi}{4}$	$\frac{1}{2}$	$\frac{\sqrt{2}}{2}$	$\frac{\sqrt{2}-\sqrt{6}}{4}$	$\frac{\sqrt{6}}{4}$	$\frac{\sqrt{6}}{4}$	$\frac{\sqrt{2}}{4}$	$\frac{\sqrt{2}}{4}$

Examples 1–2: Formulas for $\sin(\alpha+\beta)$ and $\cos(\alpha+\beta)$

1. One conjecture is that the formula for the sine of the sum of two numbers is $\sin(\alpha+\beta) = \sin(\alpha)\cos(\beta) + \cos(\alpha)\sin(\beta)$. The proof can be a little long, but it is fairly straightforward. We will prove only the case when the two numbers are positive, and their sum is less than $\frac{\pi}{2}$.

   a. Let α and β be positive real numbers such that $0 < \alpha + \beta < \frac{\pi}{2}$.

   b. Construct rectangle $MNOP$ such that $PR = 1$, $m\angle PQR = 90°$, $m\angle RPQ = \beta$, and $m\angle QPM = \alpha$. See the figure on the right.

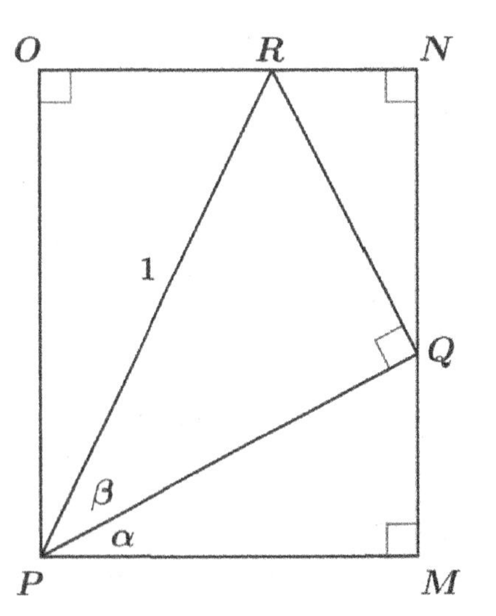

   c. Fill in the blanks in terms of α and β:

      i. $m\angle RPO =$ ____________________.

      ii. $m\angle PRO =$ ____________________.

      iii. Therefore, $\sin(\alpha+\beta) = PO$.

      iv. $RQ = \sin($_____$)$.

      v. $PQ = \cos($_____$)$.

   d. Let's label the angle and length measurements as shown.

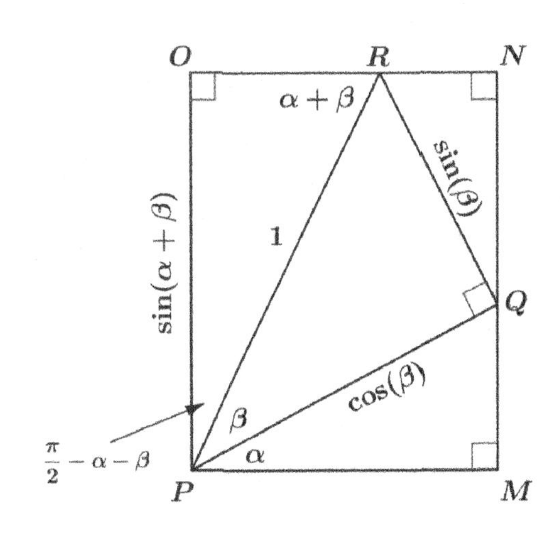

   e. Use this new figure to fill in the blanks in terms of α and β:

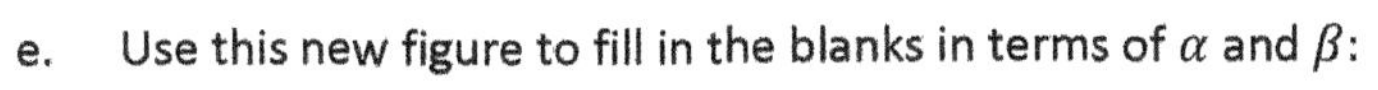

      i. Why does $\sin(\alpha) = \frac{MQ}{\cos(\beta)}$?

      ii. Therefore, $MQ =$ __________.

      iii. $m\angle RQN =$ ____________.

   f. Now, consider $\triangle RQN$. Since $\cos(\alpha) = \frac{QN}{\sin(\beta)}$,

      $QN =$ ________________.

   g. Label these lengths and angle measurements in the figure.

h. Since $MNOP$ is a rectangle, $OP = MQ + QN$.

i. Thus, $\sin(\alpha + \beta) = \sin(\alpha)\cos(\beta) + \cos(\alpha)\sin(\beta)$.

Note that we have only proven the formula for the sine of the sum of two real numbers α and β in the case where $0 < \alpha + \beta < \frac{\pi}{2}$. A proof for all real numbers α and β breaks down into cases that are proven similarly to the case we have just seen. Although we are omitting the full proof, this formula holds for all real numbers α and β.

For any real numbers α and β,

$$\sin(\alpha + \beta) = \sin(\alpha)\cos(\beta) + \cos(\alpha)\sin(\beta).$$

2. Now, let's prove our other conjecture, which is that the formula for the cosine of the sum of two numbers is

$$\cos(\alpha + \beta) = \cos(\alpha)\cos(\beta) - \sin(\alpha)\sin(\beta).$$

Again, we will prove only the case when the two numbers are positive, and their sum is less than $\frac{\pi}{2}$. This time, we will use the sine addition formula and identities from previous lessons instead of working through a geometric proof.

Fill in the blanks in terms of α and β:

Let α and β be any real numbers. Then,

$$\cos(\alpha + \beta) = \sin\left(\frac{\pi}{2} - (\underline{\qquad\qquad})\right)$$

$$= \sin((\underline{\qquad\qquad}) - \beta)$$

$$= \sin((\underline{\qquad\qquad}) + (-\beta))$$

$$= \sin(\underline{\qquad\qquad})\cos(-\beta) + \cos(\underline{\qquad\qquad})\sin(-\beta)$$

$$= \cos(\alpha)\cos(-\beta) + \sin(\alpha)\sin(-\beta)$$

$$= \cos(\alpha)\cos(\beta) - \sin(\alpha)\sin(\beta).$$

For all real numbers α and β,

$$\cos(\alpha + \beta) = \cos(\alpha)\cos(\beta) - \sin(\alpha)\sin(\beta).$$

Exercises 1–2: Formulas for $\sin(\alpha - \beta)$ and $\cos(\alpha - \beta)$

1. Rewrite the expression $\sin(\alpha - \beta)$ as $\sin(\alpha + (-\beta))$. Use the rewritten form to find a formula for the sine of the difference of two angles, recalling that the sine is an odd function.

2. Now, use the same idea to find a formula for the cosine of the difference of two angles. Recall that the cosine is an even function.

For all real numbers α and β,

$$\sin(\alpha - \beta) = \sin(\alpha)\cos(\beta) - \cos(\alpha)\sin(\beta), \text{and}$$

$$\cos(\alpha - \beta) = \cos(\alpha)\cos(\beta) + \sin(\alpha)\sin(\beta).$$

Exercises 3–5

3. Derive a formula for $\tan(\alpha + \beta)$ in terms of $\tan(\alpha)$ and $\tan(\beta)$, where all of the expressions are defined.

   Hint: Use the addition formulas for sine and cosine.

4. Derive a formula for $\sin(2u)$ in terms of $\sin(u)$ and $\cos(u)$ for all real numbers u.

5. Derive a formula for $\cos(2u)$ in terms of $\sin(u)$ and $\cos(u)$ for all real numbers u.

Problem Set

1. Prove the formula

   $\cos(\alpha+\beta) = \cos(\alpha)\cos(\beta) - \sin(\alpha)\sin(\beta)$ for $0 < \alpha+\beta < \frac{\pi}{2}$

   using the rectangle $MNOP$ in the figure on the right and calculating PM, RN, and RO in terms of α and β.

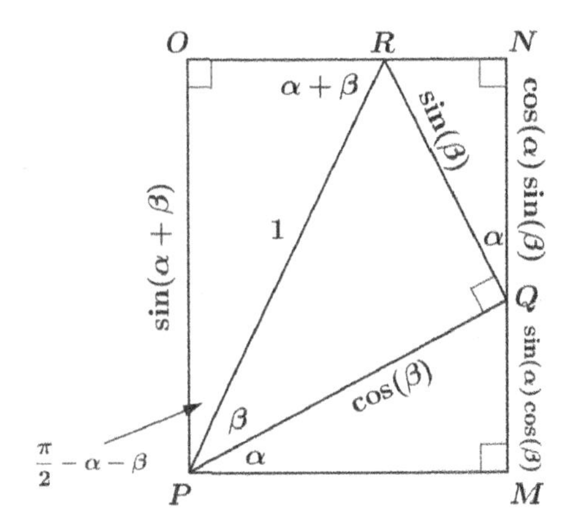

2. Derive a formula for $\tan(2u)$ for $u \neq \frac{\pi}{4} + \frac{k\pi}{2}$ and $u \neq \frac{\pi}{2} + k\pi$, for all integers k.

3. Prove that $\cos(2u) = 2\cos^2(u) - 1$ for any real number u.

4. Prove that $\frac{1}{\cos(x)} - \cos(x) = \sin(x) \cdot \tan(x)$ for $x \neq \frac{\pi}{2} + k\pi$, for all integers k.

5. Write as a single term: $\cos\left(\frac{\pi}{4} + \theta\right) + \cos\left(\frac{\pi}{4} - \theta\right)$.

6. Write as a single term: $\sin(25°)\cos(10°) - \cos(25°)\sin(10°)$.

7. Write as a single term: $\cos(2x)\cos(x) + \sin(2x)\sin(x)$.

8. Write as a single term: $\frac{\sin(\alpha+\beta)+\sin(\alpha-\beta)}{\cos(\alpha)\cos(\beta)}$, where $\cos(\alpha) \neq 0$ and $\cos(\beta) \neq 0$.

9. Prove that $\cos\left(\frac{3\pi}{2} + \theta\right) = \sin(\theta)$ for all values of θ.

10. Prove that $\cos(\pi - \theta) = -\cos(\theta)$ for all values of θ.

This page intentionally left blank